Hege Hernæs

Building the Mallaig Railway

— a photographer's story —

All rights reserved

© Glenfinnan Station Museum

ISBN 978-1-5272-7341-2

Written, researched and designed by Hege Hernæs
Printed and bound by Short Run Press Ltd., Exeter, Devon

Published by Glenfinnan Station Museum 2020

Acknowledgements

Writing this book has been a lengthy process. I make my living as a freelance translator and in my spare time, I provide curatorial services for Glenfinnan Station Museum as a volunteer. Researching and writing had to be fitted in with many other demands on my time. I would not have been able to complete this work without the help of a great many people.

A fantastic group of friends volunteered to lend a hand, and I am grateful to Gry Heggli and Miriam Iorwerth who both generously gave me the boltholes I needed to focus on my writing. My thanks also go to my band of sharp-eyed proof readers, John McGregor, Carole Hognestad and Charles Mutter. From their respective areas of expertise, they each brought invaluable input to my draft manuscript.

Special thanks go to Dr. Chris Robinson, who donated the enamel 'Railway Stores' sign to the Museum back in 2013, and to my 'neighbour', Bridget Willoughby. Without Chris's donation, the photographs featured in this book would not have found their way to us. And without Bridget's keen interest in the navvies of the Mallaig Extension and detailed local knowledge of the Lochailort topography, I would not have been able to identify some of the photographed sites that have proved pivotal to my research.

My thanks also go to a great many helpful curators and archivists at museums and archive centres around the world. Most prominent among them are the Association of Lighthouse Keepers, Arisaig Land Sea and Islands Centre, Ballyshannon and District Museum, Bute Museum, the Industrial Locomotive Society, North British Study Group, Maine Maritime Museum, the Norwegian Railway Museum, and the Royal College of Physicians & Surgeons of Glasgow.

However, special thanks go to my husband John, who created Glenfinnan Station Museum before I arrived from my native Norway. It was his gentle nudge that gave me the courage to embark on this project back in 2017. Since then he has patiently accepted my many hours of trawling round the nation's archives, weeks away on writing breaks, and constant demands on his time to help me deliberate, rethink, change tack, fine-tune. Being able to share my excitement with him every time I made a discovery that took my story forward or deepened my understanding, made my work on this book feel like an enjoyable treasure hunt.

My hope is that you will share this enjoyment, and that you will let me know should you hold any of the pieces that are still missing from the jigsaw puzzle.

Hege Hernæs,
Glenfinnan, 2020

Foreword

I'm delighted to have been asked to compose a Foreword for this significant piece of Scottish railway history. My late father Sir William was far better known throughout railway heritage circles than me but, having helped divert the East Coast mainline when I was a teenager and, given the significance to our corporate history of the West Highland Line, I'm pleased to follow on in his stead.

Last year marked the 150th anniversary of the company's formation and during that sesquicentennial year many of our staff, including myself, my wife & my eldest son walked across Rannoch Moor in the footsteps of Sir Robert, our founder. Fortunately, we performed this feat in fair weather during May (unlike Robert, who did so in January), but our walk did serve to demonstrate the magnitude of the task at hand and the hostile environment in which those involved in the enterprise were operating - a feat often lost on those travelling the route in the comfort of their railway carriage. Sadly, the main West Highland Line contract didn't come Robert's way but the research doubtless proved vital in planning the firm's involvement in the later enterprise.

The Mallaig Extension to the West Highland Line faced far greater challenges, given the mountainous terrain and the task of navigating a path through the rock so prevalent in those parts. Thomas (later Sir Malcolm) was typical of his father's offspring and similarly pioneering in his conduct - as demonstrated by the unfortunate accident that befell him. His survival and the subsequent acts that he performed in later life are still proudly remembered today by his 100 year old youngest son, grandsons & great grandsons many of whom are actively involved in the firm today.

The words that follow provide a great illustration of the determination that the Victorians had to push the railway to its geographic limits and offer access to the far distant parts of our nation.

Please share my enjoyment in this most excellently researched fascinating story.

Sir Andrew McAlpine (7th Baronet)
21 September, 2020

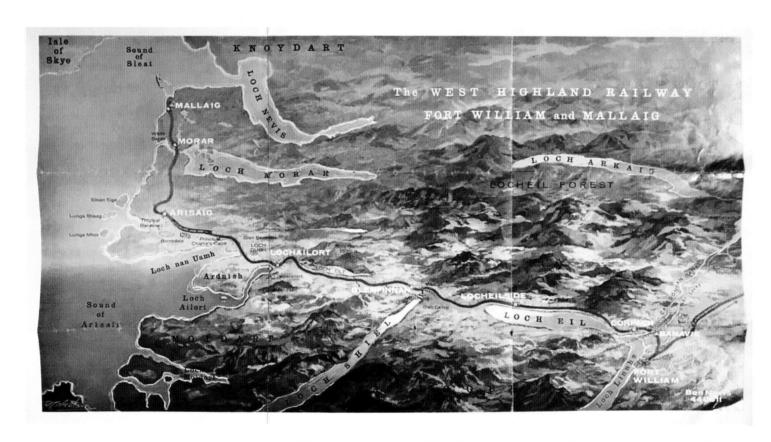

Window Gazer map, British Railways c. 1960 – Glenfinnan Station Museum collection

Contents

Introduction:	The Finding of a Photographic Treasure	1
Chapter 1:	A Railway West of Fort William
Lochaber before the Mallaig Railway – a rough sketch	5
Chapter 2:	The Men who Built the Mallaig Railway
And the folk who helped them	11
Chapter 3:	The Power of Water, the Strength of Concrete
How the Mallaig Railway was built	35
Chapter 4:	The Person behind the Camera
A photograph is a point of view	71
Appendix 1:	Maps	101
Appendix 2:	The Locomotives	111
Appendix 3:	The Turbines	115
Appendix 4:	The National Railway Museum's Photos	117
Appendix 5:	The Holden Collection as thumbnails	119
Bibliography		129

'The line commences by a junction with the existing single line branch from Fort William to Banavie. After crossing the Caledonian Canal, the new line heads generally in a westerly direction as far as Arisaig. Thence turning north, it keeps parallel with the coast line and terminates in a new port at Mallaig. The line has, as its main 'object', the development of the fishing industry on the west coast of Inverness-shire. It will undoubtedly, during the summer months, also form an attractive route to holiday makers.'

Major Pringle, Government Inspector, 18 March 1901

www.glenfinnanstationmuseum.co.uk

The Finding of a Photographic Treasure

Ours is a small independent museum tucked away amidst the mountains, moors and lochs of the Scottish West Highlands. It is dedicated to telling the story of the West Highland Line and its Mallaig Extension. Housed in the picturesque McAlpine-built Glenfinnan Station, the museum offers a hands-on introduction to the area's railway heritage and an up-close view of the Jacobite steam trains as they pause at the station on the journey between Fort William and Mallaig.

In the summer of 2017, the museum received an email with a photo attachment from a Michael Holden. As the museum's curator, I was amazed to open the file and find a black and white photograph of a ramshackle building carrying an enamel sign that read 'Cooper & Co's Railway Stores'. Michael had found a picture of a similar but battered sign on the museum's website. The battered sign is indeed part of our collection, but its image appears as a banner for our online shop. Michael Holden thought we might find it interesting to see the sign in its original context.

Despite having spent a good chunk of my life researching the history of the Mallaig Extension, I had never come across a picture of any of the twelve Cooper & Co. railway stores that were in operation along the line during its construction. While I had a good idea of their general whereabouts, there were, to my knowledge, no records that could tell me what they looked like or where their precise locations were. Suddenly, thanks to a picture received from out of the blue, I knew the exact nature of a railway store and had a means of accurately pinpointing where one of them had stood. For while much has changed along the Mallaig Extension over the last century, the outline of the mountains remains the same. The importance of the picture dumbfounded us at the museum, and we were more than a little excited. However, it turned out that this was only the first of many jaw-dropping moments to come.

On thanking Michael Holden for sending through the photograph, we tried to convey its historical significance to him. We also asked how it came to be in his possession. He responded by return: he had bought, on the off-chance, a batch of unlabelled cellulose nitrate negatives from an auction house in Cornwall.

Cellulose nitrate is a form of plastic that was commonly used for photographic materials in the period between the late 1880s and the early 1930s. As nitrate-based negatives age, they become prone to spontaneous combustion and need to be stored in the right environmental conditions to be kept safe. The fact that this unlabelled batch had been bought by someone as knowledgeable and passionate about historic photographs as Michael Holden, was a stroke of pure good fortune.

On processing the fragile pictures, he had recognised an inscription that made him suspect the many railway-related scenes could well be featuring the building of the West Highland Line. From there, the internet had taken him to our website. The batch he had bought contained some 240 photographs! Over the course of the next few days, he generously sent low-resolution scans of them all through to the museum. To us, what we received was nothing short of a treasure.

Overleaf: Cooper & Co.'s railway stores at Kinlochailort (HC093).

Subjects

Approximately 150 of the photographs clearly featured scenes from the building of the Mallaig Extension to the West Highland Line, or more specifically, the construction of the section between Glenfinnan and Mallaig.

Mixed in with these West Highland photos was a batch of scenes we were unable to identify locally. These images looked like present-day holiday snaps: impressive buildings, scenic waterfalls, beautiful bridges and sailing boats. The variety of subjects puzzled and intrigued us. Michael Holden believed that the photographs had been taken with the same sophisticated camera. However, the order in which the pictures were taken was unfortunately impossible to establish.

Initial dating

The construction of the Mallaig Extension began in January 1897 and was completed at the end of March 1901. The pictures clearly showed the works towards the end of the contract period. Our initial approximate dating of the photos was therefore 1899–1901.

Copyright issues

Michael Holden requested the museum's help in identifying as many as possible of the subjects and locations featured, in the hope of detecting the identity of the photographer. According to current legislation, the copyright belongs to the heirs of a photograph's originator for a period of 70 years after his/her death, and Michael was keen to have the legal issues sorted out.

We therefore set about looking for the people and places photographed by our mystery camera operator, hoping to prove who had been behind the lens. What follows is the story of this search, and the many things we learnt along the way, about the building of the Mallaig Extension and the many people involved.

Towards the end of the research period, we did however discover that because the photos were taken before 1 June 1957, they were covered by earlier copyright legislation which provided protection only for a period of 50 years from the year they were taken. According to the new rules, this period would have been extended had the photos still been in copyright in July 1995, but this was clearly not the case here.

Sources

Many good books have been written about the building of the Mallaig Extension and its contractor, 'Concrete Bob McAlpine', perhaps most famously by John Thomas, whose first edition of *The West Highland Railway* was published in 1965. The indefatigable Sir William McAlpine (1936–2018) secured John Thomas an interview with Concrete Bob's third son, Sir Malcolm McAlpine (1877–1967), while he was writing his book. Many of the Mallaig Extension stories featured by John Thomas are therefore based on Sir Malcolm's first-hand account.

A more detailed description of the practicalities involved is given by Iain F. Russell in his *Sir Robert McAlpine & Sons – the early years*, which is largely based on documentary evidence from the company archives, currently held by the University of Glasgow Archive Services. The archive services also hold the Engineer's Certificates for the building of the Mallaig Extension. These certificates provide detailed information about the work carried out by the Contractor and paid for by the Engineers. A detailed study of these paybills has formed a rich source of construction detail for our Chapter 3.

Two further unpublished biographies of the McAlpine family have been written; one by Alec M Hardie in c. 1960, another by Compton MacKenzie (of Whisky Galore fame) in c. 1940. The typewritten manuscripts are held by the University of Glasgow Archive Service and by the National Library of Scotland respectively. The former is based on interviews with family members and employees, the latter on the first-hand account of Robert McGregor, later to become a famous civil engineer in his own right. At the time of the Mallaig

Introduction: The Finding of a Photographic Treasure

Extension construction, McGregor was however a mere youngster, just 14 years old when he first joined the navvies with his ganger father in 1897. According to Robert McGregor himself, he could 'handle practically any tool like a fully-grown man'. However, while his account is full of colourful social detail, his recollections are understandably less reliable when it comes to numerical accuracy.

Historian Dr John McGregor's books provide the most authoritative account of the political background to the West Highland Railway and its Mallaig Extension. They give an insight into the many opposing interests and conflicting ideas that influenced and shaped events, with details of all the personalities whose work, thoughts and financial interests had a bearing on the outcomes.

For engineering detail, the proceedings of the Institution of Civil Engineers include specific papers on the West Highland Line by both Walter Stuart Wilson and James Shipway. Engineering journals published at the time also give good accounts of the works. For background information on concrete in the late 19th/early 20th century, H.G. Tyrell's *Concrete Bridges and Culverts, for Both Railroads and Highways* from 1909 is a fascinating read. The same goes for John Haldane's *Railway Engineering*, published in 1908, with respect to general engineering developments.

The National Archives in Kew hold a wealth of official correspondence and reports relating to the construction of the Mallaig Railway. Combined with the British Library's collection of newspapers, these sources provide a multifaceted layer of background detail and reliably dated records of important events.

The National Railway Museum's three albums of photographs associated with the construction of the Mallaig Extension are also sources of important information. Three of these pictures have been included in our Appendix 4 under licence from NRM/Science and Society Picture Library. All other photographs reproduced in the following chapters form part of Glenfinnan Station Museum's 'Holden Collection' and a collection-specific reference is provided for each image.

While all these records and publications have informed our detective work, the most important source of specific detail and context has been the local police records for Glenfinnan and Kinlochailort[1], held by Lochaber Archive Centre in Fort William. These cover the period between April 1897 and March 1901 and provide reliable first-hand eye-witness accounts of life along the Mallaig Extension during the construction period, and they are rich in ethnographic detail.

We are deeply indebted to the late Sir William McAlpine (1936–2018) for his enthusiastic assistance. As the great grandson of 'Concrete Bob', Sir William had supported the museum's work since it was first established. We therefore shared all the pictures with him soon after they came to light in 2017, and in response, he promptly gave us access to family photographs to aid our research. He also gave generously of his time, allowing extensive interviews over the phone, thereby providing inimitable tales from the building of the Mallaig Extension, passed on to him from his great uncle Malcolm, who he had been very close to. We are grateful to Sir William's widow, Lady Judy, and his son, Sir Andrew, for their continuing support of our work.

---

[1] 'Lochailort' didn't become a place name until 1901, when the railway station was given its name. Professor Blackburn of Roshven objected to no avail: 'Kinlochailort is a place; Lochailort is a loch!'

1 A Railway West of Fort William

Lochaber before the Mallaig Railway – a rough sketch

The original West Highland Line was built over a period of five years (1889–1894) towards the end of the great railway-building era in British history. The remote single-track railway line from Craigendoran (just north of Glasgow) to Fort William was constructed after much political and commercial jostling. While the West Highland Railway was intended to provide a transport link to the south for the impoverished crofting population on the west coast, it proved impossible to agree on a route for the line west of Fort William before Parliament was scheduled to authorise the building of the railway in 1888/89. Much to the delight of the Fort William business community, work nevertheless went ahead with Fort William as the temporary terminus.

Parliamentary approval for the line west of Fort William materialised only a week before the town celebrated the opening of its railway to Glasgow in August 1894. The guarantee that was required for construction to go ahead, was also eventually approved, but not until a full two years later, with a short branch to Banavie Pier on the Caledonian Canal completed in 1895, as another interim railhead for all the districts to the west.

The idea that it was in the national interest to build a railway to the coastal area between Oban and Strome Ferry was launched in the Napier Commission's 1884 report on the public inquiry into the condition of crofters and cottars in the Highlands and Islands. Lord Francis Napier, a diplomat and colonial administrator of Scottish origin, had been asked to chair the inquiry in response to the 'Crofters' War' of the 1880s. The Victorian elite in the south had been shocked to hear of armed skirmishes between British soldiers, acting to defend the interests of landowners, and crofters who were distressed by high rents and the threat of eviction. The response from the ruling classes was to incentivise the building of transport infrastructure, to give Highlanders and Islanders a means of getting their fish to markets in the south and enjoying regular supplies of salt, barrels, itinerant packers etc. The hope was that this would prevent or at least delay fundamental land reform.

The path to political agreement about the building of an extension to the west was a long and arduous one. The many conflicts that had to be resolved in order eventually to give the coastal population the railway they needed have been covered in detail by Dr John McGregor in his books on the West Highland Line. In our context, the important thing to note is that one of the conditions the promoters of the line had to fulfil in order to receive the guarantees and subsidies they needed, was to build a breakwater and pier on the foreshore at Mallaig, which until the arrival of the railway had been 'just a pretty wee bay'[1].

---

[1] Donald McDonald of the Highland Railway in his testimony to Parliament, 1894

The local population

Fort William

In the early 1890s, Fort William was a small town with a population of just under 2,000. The arrival of the railway in 1894 created a need to increase the local housing stock, and the operator, the North British Railway Company, had been forced to build new tenements for its incoming staff. The new railway had also changed the layout of the town, curtailing public access to the foreshore where local fishermen previously used to moor their boats and dry their nets. This caused significant local tension (McGregor, 2015).

While many made a living from the herring fisheries, there was also a growing middle class. The town had boasted a hospital since 1865, and there was an enterprising group of businessmen who regularly advertised their wares in the local newspaper, the *Oban Times*, which covered the entire area of the then Argyll-shire. There were advertisers based in Oban and Glasgow who sought to sell such middle class treats as artificial teeth, pain free dental services, herbal cures, sheet music, insurance, pianos and organs. Dancing lessons and dressmaking classes were advertised locally in Fort William, and there was a thriving amateur dramatics society. Local seed merchants Marshall and Pearson regularly marketed their services in the paper. Adverts for whisky, ale, rabbits and poultry frequently recurred, as did the Ben Nevis Auction Mart's sales of horses and livestock.

The *Oban Times* carried all the local news from within its area, such as reports of concerts, social events, accidents, contracts awarded and completed, deaths and marriages, shinty matches and Highland gatherings. A regular poetry column occasionally featured works in Gaelic. In the period 1899–1902, news of the Boer War became an important feature, since many of the area's able-bodied men had joined the local regiments to fight for British economic interests in South Africa.

The *Oban Times* had set up its own 'Relief Fund for the Widows and Orphans, Relatives and Wounded of our Highland Soldiers and Sailors', taking care to stress that this was not a charitable initiative, but a reward for the military service of loved ones.

The rural areas to the west

Parliamentary reports tell us that the estimated population in the predominantly Gaelic-speaking area between Glenfinnan and Knoydart was 1,790 in 1894, while the number of residential houses stood at 381.

According to the County Medical Officer's report to Inverness-shire County Council in 1891, Lochaber stood out from the rest of the county:

> [Lochaber] differs from all the other districts in the County of Inverness. The District Committee have not appointed a District Medical Officer, and they have not adopted the Infectious Disease Notification Act. I have, therefore, been unable to obtain accurate information as to the state of the district. I have visited some of the most insanitary portions of it, and have observed more insanitary surroundings than I have seen in any other part of the mainland (Ogilvie, p. 18).

He voices his concern with the general living conditions in no uncertain terms:

> A remedy is urgently needed. If this state of matters existed in some remote colony, probably attention would have been given to it long ago […] above all, we have to deal with an essentially poor people, who yet, owing to their remoteness from markets, have had to pay high prices for the most ordinary necessaries of life (Ogilvie, p. 14).

There were schoolhouses in villages like Glenfinnan, Glenuig and Arisaig, with so-called side-schools in smaller settlements. Schooling was compulsory for all children between the ages of 5 and 13 thanks to the Education (Scotland) Act of 1872, but according to school records, attendance was often low due to sickness, heavy rain, snow, or 'cutting of the peat'.

In Compton MacKenzie's unpublished biography of the McAlpine family, Robert McGregor describes the topography of the area in these terms:

> 'There were streams, lochs, ravines, sheer faces of rock, and much bog. The sea lochs penetrated inland with rocky almost impassable tongues of land dividing them. [...] the only road was very narrow and almost its only traffic was the daily stage-coach' (MacKenzie, p. 3/2).

Loch Beag and Loch nan Uamh between Lochailort and Beasdale (HC076).

1: A Railway West of Fort William

The road from Fort William to Arisaig had been built by Thomas Telford in the period between 1803 and 1812 as the first ever British infrastructure project that was awarded a public subsidy. However, the state of the road made a journey by horse and carriage rather uncomfortable, and the 36-mile trek took 7 hours.

At that time, it was customary for most people to walk vast distances on a regular basis. Although bicycles with even-sized wheels (introduced in the 1880s) were becoming increasingly popular, they remained expensive and essentially ill-suited for the bumpy road. Horses and carriages were exclusive to the rich. People of meagre means – i.e. most people – would either row or walk if they needed to get anywhere to attend school, social gatherings, religious services and the like.

The vast swathes of mountainous land west of Corpach were characterised by the huge estates of wealthy landowners in grand mansions. They employed local staff as estate managers, shepherds, horse handlers, gillies, gardeners, gamekeepers, cooks, domestic servants etc., but their land was largely tended by crofters and cottars[2] with no security of tenure. Tenants sometimes faced hard toil and desperately impoverished conditions while struggling to keep hunger at bay. Emigration appeared to be the only solution to many who left the shores of Scotland for Canada, the USA or Australia.

While some landowners were keen to help alleviate the plight of the local population by promoting the railway, others were more interested in protecting the privacy of their estate, their income from hunting and fishing, and the scenic beauty to which they felt entitled.

Most landowners accepted the payment of a lump sum from the railway company as compensation for their loss of 'amenities'. Mrs Head of Inverailort Estate decided instead to file a Parliamentary petition opposing the West Highland Railway (Mallaig Extension) Bill:

> 'It appears ... that the railway is so laid out to pass right in front of the mansion house and to intersect the property in the most injurious and destructive manner by cuttings, viaducts and tunnels almost from end to end, entirely destroying the amenities of the mansion house by passing in full view of the windows, and your Petitioner objects thereto, and confidently alleges that no case of public necessity or advantage can be made out in favour of the line at all commensurate with the wanton injury which would be inflicted upon the property, rights, and interests of your Petitioner and other landowners affected by the Bill.' (Petition against the West Highland Railway (Mallaig Extension) Bill 1894, s. 11, p. 4.)

While we may well find the Heads' concerns less than community-spirited, they were clearly right to fear that the railway would bring radical change. For many, that change was of course sorely needed.

---

[2] Crofters were tenant farmers who paid rent in kind for their house, a plot of arable land and a right of pasture held in common with other crofters. Cottars gave their labour for the right to live in a cottage – there was no land attached.

Opposite: The daily mail cart on the narrow road from Fort William to Arisaig, here heading west from Glenfinnan to Loch Eilt (HC212).

2 The Men who Built the Mallaig Railway

and the folk who helped them

The engineers for the railway line from Craigendoran (just north of Glasgow) to Fort William had been Formans & McCall, while Lucas & Aird had been the contractors. Both of these Glasgow-based firms had in 1894 been expected to win the contract for the Mallaig Extension as well. But when Parliament rejected the proposed Mallaig Railway Guarantee Bill that year, Lucas & Aird gave up the waiting game and upped sticks. They dismantled their Corpach base camp and took their equipment and navvies south to pastures new. Instead, when the Guarantee Bill eventually did go through in August 1896, the contracts for the new railway went to two other Glasgow-based firms: engineers Simpson & Wilson and contractors Robert McAlpine & Sons. The two firms had worked alongside one another on a section of the Glasgow underground in 1892–94, experimenting with concrete as a material for lining the tunnels (Russell, p. 6).

Alexander Simpson and Walter Stuart Wilson

The Simpson & Wilson civil engineering partnership was founded by Alexander Simpson in 1878. They specialised in railway work and particularly tunnelling for the North British Railway (NBR), for which Simpson was later to become a director.

Opposite: Young girl at the Polnish shelter for railway engineers (HC094).

Alexander Simpson (1832–1926) first gained prominence as the engineer of a railway in the Dominican Republic, financed by Glasgow investors. On his return to Scotland in the early 1880s he was appointed engineer to the Glasgow and City District Railway Company, a subsidiary of the NBR. The Glasgow Subway (built 1891–1896), the third underground railway in the world after London and Budapest, was Alexander Simpson's brainchild.

Walter Stuart Wilson (1850–1926) first worked with Alexander Simpson in 1874–77 on the Glasgow, Bothwell, Hamilton & Coatbridge Railway where Wilson was the resident engineer, carrying out all the work under Simpson's supervision. He must have done a good job as the distinguished Simpson soon offered partnership to the 18-years-younger Wilson.

Robert McAlpine & Sons

Robert McAlpine set up his first construction company in 1868 at the age of 21, possibly to impress his prospective parents-in-law, who did not approve of his liaison with their daughter Agnes. The pair were Lanarkshire Sunday school teachers, but Agnes was the daughter of a master bootmaker while Robert was a lowly bricklayer. In the course of the next couple of years however, Robert married Agnes and became father to Robert jnr. while working himself up from a jobbing tradesman to a building contractor. A decade later he had started gaining a reputation for building tenements in blocks of concrete (*Hamilton Advertiser*, 13 May 1876) (Russell, p. 21). By 1890, he had already been using mass concrete extensively in the building of the Lanarkshire & Ayrshire

'The Bungalow' in Corpach, where Robert McAlpine jnr. and his wife Lilleas moved in as newly-weds in January 1897. The building still stands, within extensive grounds (HC089).

Railway, much encouraged by the Engineer-in-Chief, John Strain (1845–1931). The new material was cheaper to use than traditional bricks and did not require the services of skilled bricklayers.

By the time the Mallaig Extension contract came along in 1897, Robert was 50 years old. He had become a widower in 1888 but met and married his second wife Florence in Dublin in 1889. His business had re-established itself after a devastating bankruptcy in 1880, and three of his five sons with Agnes were working in the company, along with many other members of his extended family.

They had several other major contracts on the go during the Mallaig Railway construction period, but although business was booming, the enormous workload gave rise to significant cash flow and manpower problems.

In 1897, Robert snr. sent his 28-year-old eldest son Robert jnr. (1868–1934) to the West Highlands to take charge of the works there. He arrived with his wife Lilleas straight from their honeymoon in France to settle in 'The Bungalow' in Corpach. When they moved back to Dunbarton-shire, their departure was said to be met with 'great regret' in the local community (*Oban Times*, 15 Sept 1900).

2: The Men who Built the Mallaig Railway

Robert senior stayed behind in the Lowlands to deal with the company's financial affairs and tendering activities. He leased a house in Old Kilpatrick with his new Irish wife, Florence, and their three children, as well as his four youngest children with Agnes. Their grand Dalnottar House was located on the banks of the Clyde estuary, near to what is now the Erskine Bridge. He was however 'always about [on the Mallaig Extension] in a carriage and a pair of horses' (MacKenzie, p. 3/4).

Robert's second son, William (1871–1951), also played a significant part in building the railway to Mallaig. In his McAlpine biography, Compton MacKenzie quotes Robert McGregor:

> 'For some reason or another I was afraid of Mr Willie. I believe this was caused by hearing the workmen talking about him so much. Everyone knew he was the last word in the firm, and even the other partners always consulted him on matters of importance. His word was final with them' (MacKenzie, p. 3/6).

This may well have been because William was in charge of the head office in Glasgow and had the job of negotiating with clients and raising money to make sure the company could deliver on its many contracts (Russell p. 85).

In April 1898, William married Margaret Bishop, the younger sister of Lilleas who had married Robert jnr. two years earlier.

Only one year earlier, in 1897, the McAlpine brothers' sister, Mary, had married the Bishop sisters' older brother, Andrew. In other words, three of the ten McAlpine children shared a father-in-law: their father's old friend, Thomas Bishop.

William Hepburn McAlpine (1871–1951), the second eldest brother, apparently photographed from a photo album (HC126).

Thomas G. Bishop

Robert jnr.'s father-in-law, Thomas Bishop (1845–1923), lived with his family in Dalmore House at Rhu near Helensburgh. After years of working for tea merchant Thomas Lipton, he set up his own grocery business in 1871. He called it Cooper & Co. after his mother-in-law, who lent him the start-up money. By the time work began on the Mallaig Extension, he had set up shops in most of the towns in western Scotland.

It comes as no surprise then, that when the McAlpines looked for ways of keeping their West Highland labour force in food, clothes and supplies, they turned to Cooper & Co. for assistance. Thomas Bishop's company set up twelve stores on the remote stretch west of Kinlocheil, where no other shops existed. They were contracted to supply the stores with their usual high-quality goods at a reasonable price. Securing enough men to push on with the works at the required speed proved to be a major problem for the contractors, and good provisions formed part of the McAlpine strategy to attract workers to their remote work sites.

Awarding the contract to Coopers & Co. may well have been a wise move. When the McAlpines ran into severe cash flow problems in the period 1900–03, Thomas Bishop provided Clydesdale Bank with the much-needed guarantees required to extend the McAlpine overdraft, to the tune of £23,000 over the period (Russell, p. 94). The entire Mallaig Extension Contract was worth only 14 times as much.

Thomas Malcolm McAlpine

Robert snr.'s third son Thomas Malcolm (1877–1967) also travelled north to work on the Mallaig Extension contract. At the age of 19 he was made assistant to his older brother Robert.

The young Robert McGregor knew Thomas Malcolm simply as 'Mr Tom' and considered him to be the most energetic man he had ever met. He was a hard

Thomas Malcolm McAlpine (1877–1867), third eldest of the McAlpine brothers (HC134).

taskmaster but very popular with the younger men, who tried to keep pace with him.

McGregor told Compton MacKenzie:

> 'It was a life that required a great deal of endurance and vitality, and it meant really hard living. For months he [Mr. Tom] occupied a small hut not more than six feet square. He said he did not mind it so much when he had it to himself but protested when asked to share' (MacKenzie, p. 3/10).

Railway contractor accommodation near Polnish (HC116).

Thomas had been working for the family business since he left school at 16. He showed a keen interest in engineering but was not given a chance to qualify as an engineer until later in life, after he had successfully supervised the building of the Leadhills and Wanlockhead Railway in Scotland and the Waterford to Rosslare Railway in Ireland. He was accepted into the Institution of Civil Engineers in 1908 (UK Civil Engineering Records, 1820–1930).

As a teenager he was clearly interested in the mechanical gadgets of his era, and was called on as a photographer by the family, verifiably from the age of seventeen:

> On Saturday 18th inst., the choir of St. Columba's Clydebank, had their annual outing on the banks of the Gareloch. In the loveliest of weather, the boys played cricket, football, etc., in the fields of Ardenconnel, to which they had been kindly invited by Mr and Mrs McAlpine. [...] In a brief interval, Mr Thomas McAlpine photographed the group [...] (*Dumbarton Herald*, 26 Sept 1894).

In later life, Thomas Malcolm (who changed his name to Malcolm in 1901) went on to become one of the most influential leaders in British construction industry. He was knighted for his World War I achievements in 1921. In 1944, he designed the concrete breakwaters for the Mulberry Harbours used in the D-day landings.

In 1930, Sir Malcolm built the all-concrete, ultra-modern, extreme-luxury Dorchester Hotel in London, which stayed in the McAlpine family's ownership until 1977.

However, when he left the family home in Old Kilpatrick to travel to the West Highlands in 1897, he was a mere youngster, full of the swagger of youth and privilege, but with an open mind and a way with people.

Navvies – digging, drilling, blasting, cutting, picking, heaving, hoisting, shovelling, ballasting …

The McAlpines needed a workforce of thousands to have a hope of completing the Mallaig Extension contract on time, and although men came from every part of Britain and beyond to work in what was considered a very remote location, the majority were from Ireland, Lewis, Skye and the Highlands. Robert McGregor told Compton Mackenzie:

> 'The Irish were most of them in the Militia or were old soldiers from regular regiments. Many of the Highlanders, too, were old soldiers, but there were many sailors from the Isles. They had followed this type of work about for the best part of their lives and no Labour Exchange had spoiled their liberty: they were as free as the air they breathed. There were quite a few […] ex-convicts, but nobody bothered about their past. All that mattered was their behaviour on the job and their ability to be good comrades' (MacKenzie, p. 3/2).

We should keep in mind that the Mallaig Extension was not the only railway being built in the West Highlands at the time. Construction of the Invergarry & Fort Augustus Railway began only a couple of months after the West Highland Railway Company started work on the Mallaig Railway, and in September 1898, the Callander & Oban Railway began work on their Ballachulish Branch.

There was therefore great competition for construction workers, both skilled and unskilled, and the number of temporary workers in the areas surrounding Fort William greatly outnumbered the permanent population.

In fact, the McAlpines were never able to attract, and retain, as many workers as they wanted, so had to make sure that their terms were competitive. Morale was often low due to swarms of midges and incessant rain in summer, freezing temperatures and harsh winds in winter. Only the heaviest of downpours would stop work, for the men were on piecework so any extended pause was unpopular and often caused them to seek work elsewhere (Russell, p. 81). Many would only stay a couple of weeks.

The railway construction workers, or navvies, were organised in teams headed by a 'ganger', or subcontractor, who was responsible for collecting and paying out the wages for each of the men in his team. They had often wandered all over the country with their favourite contractor or ganger, specialising in their own particular job, as tunneller, timberman, horse handler, concretor or platelayer (Hardie, p. 34). On the Mallaig Extension, the standard pay for navvies was up to 5d per hour, or just under 4 shillings per 10-hour day, while gangers would earn a shilling more. This basic wage matched the Scottish national average for labourers in 1900 (Russel, p. 88), but there was also a small 'payment by result' bonus. The working day started at 7am sharp out on site and pay would be deducted for those who failed to show up on time.

Workers were accommodated in large wooden-framework huts made up of three compartments. The largest compartment provided sleeping quarters for up to forty men:

> [McGregor to Mackenzie:] 'The bunks consisted of a number of posts along each side of the room about four feet from the wall, plus one continuous shelf that ran the whole length, three feet from the floor. Four feet above this was another shelf of some kind. The shelving was divided by partitions into sleeping compartments in each of which two men slept. None of them wore any kind of night garment; they slept in their shirts. There were no washing sheds; there was nothing except a few basins out in the open and the nearest peaty burn' (MacKenzie, p. 3/9).

The second largest compartment was for cooking and eating:

> [McGregor to Mackenzie:] 'Men came into this room straight from work, grimed, muddy and usually wet. One central stove for cooking with a large heated surface known as a hot plate upon which every man cooked for himself. The drying of wet clothes was effected by standing as near as possible to the cooking stove' (MacKenzie, p. 3/9).

The third compartment was the hut keeper's partition, where he would live with his family or an assistant.

Ballasting just west of Borrodale Viaduct (HC233).

2: The Men who Built the Mallaig Railway

Navvies were accommodated in wooden framework buildings. Engineers and clerical staff lived in wooden or corrugated iron structures similar to those of the navvies, but with separate cubicles for each person (HC231).

Clusters of such huts formed navvy camps at various points along the entire length of the line from Corpach to Mallaig. The largest camp was in Lochailort, where a total of some 35 huts housed as many as 1,500 navvies at the peak of activities, i.e. a settlement almost the size of Fort William. This was bound to cause grievances amongst the local population, and Compton MacKenzie reports on a fierce sermon by a parish priest: 'I sometimes wonder whether this railway is being made to Mallaig or to Hell!'

Hut 'foundations' – wooden planks in the ground.
This allowed space for keeping hens below the floor boards.
It also made it easy to move huts between sites as required without an environmental impact. There is no trace of these buildings today (HC117).

McGregor makes no mention of toilet facilities when describing the hut layout. Local crofter Ewan Stewart does, however, mention the lavatorial situation in a complaint he made to PC MacGruer in May 1900. Someone was habitually using his sheepfold to relieve themselves, rendering it impossible to use for its intended purpose.

Ewan pointed the finger at James Carraig, the keeper of the navvy hut forty yards away from his croft:

> 29 May, 1900: [...] in his hut he keeps thirty or forty men [...]. There is no privy near the hut in question and although I have caught no person in my sheepfold, I have no doubt whatever but it is the men resident in James Carraig's navvy hut who use it as a privy. I want no person prosecuted but I want to get the nuisance in the fold removed and people prevented from entering and using it as a privy. (PC MacGruer, Glenfinnan)

Large sheepfold, dry stone walls topped with turf (HC216).

'there were no washing sheds'...

We have seen that Robert McGregor talked about the lack of washing sheds. Sadly, this lack of sanitary facilities in the navvy camps did not only cause disease, discomfort and unimaginable odours. Bathing in the great outdoors came with its own risks:

> FATAL ACCIDENT – On 14 June 1899, 19-year-old Michael Sullivan, an unmarried joiner, drowned in the river at Bréin-choille, just west of Glenfinnan, while bathing.
>
> Anthony Finnigan explained to PC MacGruer how it happened: four young lads, all staying in the same hut, had set off for a dip in the river near the camp. When they had found a suitable bank, two of them, both confident swimmers, stripped off immediately and swam to the opposite bank and back. Michael also stripped naked, but was still standing on the bank with his mate Patrick when the two confident swimmers set off down the river.
>
> After about forty yards they heard a shout, and turning round, they saw Michael with uplifted hands disappear in the centre of the river. They immediately swam back up to assist, but when they got there, there was no sign of Michael. Patrick, however, had jumped into the water to come to Michael's rescue when he saw him sinking, but he had done so with his boots and clothes still on, and was now in distress in the water. The two confident swimmers managed to save Patrick, but Michael had already drowned.

> FATAL ACCIDENT – In the afternoon of 31 July 1897, Duncan Gillies, a 39-year-old foreman blacksmith staying at the Camas Driseach camp, drowned 30 yards from the shore while bathing.
>
> According to PC Mackay at Kinlochailort, it was the foreman joiner at Camas Driseach who went out in a boat to pick up Duncan's body. He was then carried to the Railway Store by PCs Mackay and MacPherson. Their next task was to notify Duncan's wife, who lived in Anniesland on Great Western Road at Maryhill in Glasgow. The couple had seven children.

Bathing in the great outdoors came with its own risks. The beach at Camas Driseach at the head of Loch Ailort (HC051).

Hut keepers and their wives – domestic orderlies

It cannot have been easy to keep the peace in a hut where 40 hardworking men were living on top of each other. However, hut keepers would often work as gangers or foremen on the railway and were ideally able to lead the men with a certain level of authority.

Hut keepers claimed rent from each of the occupants, but in turn paid rent to McAlpine & Sons for the hut. They also bought coal, cooking utensils and bed clothes from the contractor. In return for the responsibility, hut keepers were afforded a modicum of privacy and could enjoy the company of a wife and children, which often put them in a position of envy. Some must have felt more than a little threatened and took measures to defend themselves:

> February 1, 1898: Alastair Macdonald (45), Hut Keeper, Scandale Bridge, Kilmallie was accused of shooting and wounding James Russel (39), Navvy residing in the accused Macdonald's hut. (PC Campbell, Glenfinnan)

In court, Macdonald maintained that the revolver was used in self-defence, as Russel had attempted to assault him and force his way into the portion of the hut occupied by his wife and children. He was let off with a £4 fine.

With a going rate of 4^d for a glass of illicit whisky, many hut keepers sought to top up their earnings, and keep in with the men, by selling unlicensed spirits:

> August 28, 1899: [...] Cautioned and charged Hugh Harvey (43), married, Hut Keeper Arienskill. Charged with Shebeening[1] to which he admitted and said that if he would not be prosecuted for this case that he would stop shebeening as long as he was in the place. (PC Sinclair, Kinlochailort)

Unsurprisingly, it appears to have been the hut keeper's wife who looked after the domestic arrangements:

> June 27, 1898: Complaint from Catherine Kennedy or MacLean, wife of William MacLean, Navvy, Glen Glenfinnan. 'George [...] and another navvy came to my hut and began to serve the stranger with the other men's tea.

---

[1] Shebeening: Illicit sale of unlicensed alcohol.

I objected to this so that they both handed to me five pence which I handed back to them saying that there was no room for them. Accused then started to curse and swear and using most abusive language. He then threw two bowls at my head [...] and afterwards tried to strike my husband, so that I had to lock the door. He then got quite outrageous and saying that he would kill both me and my husband before morning. He struck at the locked door with his boots [...] and broke the centre of the door. When the constable was removing him from the hut, he threw a bread loaf at my husband [...].' (PC MacGruer, Glenfinnan)

The lot of a hut keeper's wife was clearly far from idyllic, and many must have felt vulnerable. The police reports include only one accusation of rape, but there will undoubtedly have been other instances that went unreported.

The most frequent complaint made by the wives of hut keepers concerns what we now refer to as domestic abuse. One such instance is of particular interest as it reflects not only on the status of women, but on the standing of the native population's Gaelic language:

> Nov 11, 1899: Complaint by Christina Cameron (36) against husband John Reid (40), Hutkeeper, Glenfinnan. 'He quarrelled with me for talking Gaelic. I remained quiet, knowing that if I was to answer him he would be sure to strike me. Although I did so, he took hold of my breast with one of his hands and pushing me back against the wall with his other hand, struck me several times on the face and nose whereby blood flew from my nose. I succeeded in releasing myself from his hold and ran outside whereupon he came to the door and threw several stones at me [...]. He locked the door and would not let me inside. I gave the accused no provocation and wish him prosecuted'. (PC MacGruer, Glenfinnan)

Engine drivers – keeping the wheels turning

As the work progressed, temporary track was laid for a contractor's railway to carry materials and men to the various work sites. Locomotive drivers enjoyed a high status and were well respected, while the fireman could be a mere 'nipper'.

Sometimes, driver and fireman would be father-and-son-teams working to boost family earnings and ensure that the father's skills were passed on to the next generation of breadwinners. (See Appendix 2, p. 111, for details of some of the locomotives used during the construction period.)

Locomotive drivers enjoyed a high status and earnt respect, while the fireman could be a mere 'nipper'. Note the kettle in the hand of the young lad. Location: Carnach, by the Brunery Burn, just east of Larachmhor Viaduct (HC035).

2: The Men who Built the Mallaig Railway

Divers – braving the waters

According to the Engineer's Certificates, the building of Mallaig Pier did not start until April 1899 and was not completed until a month before the line was opened. The job required the services of divers to clear away seaweed, clean rock under water and work with the crane operators to accurately position the pre-cast concrete blocks.

The picture below shows that these brave men were wearing standard diving dress, consisting of a metal diving helmet, a waterproof canvas suit, a diving knife and weights to counteract buoyancy. Air was supplied from the boat by a pump that was manually operated by a large wheel, seen here under the boat's canvas.

Divers in Mallaig harbour (HC012).

White-collar workers – keeping tally

To build a railway, the contractor also needed white-collar workers on site, to make calculations, check on progress, pay out wages, keep a tally, take decisions. Engineers and clerical staff such as cashiers and timekeepers lived in wooden or corrugated iron-clad structures similar to those of the navvies, but with separate cubicles for each person. Local inns and farms were also used to house the contractor's most high-ranking staff, and Simpson & Wilson's engineering staff, which must have been welcome revenue for local innkeepers.

Mallaig hostelry. The sign above the door reads 'Glasnacardoch Hotel, McLellan' (HC100).

White-collar worker at Polnish schoolhouse camp, possibly Patrick Rooney, a 28-year-old cashier who stayed there (HC138).

Storekeepers – purveyors of allsorts

The men who looked after the provision stores along the route, the storekeepers, were all employed by Thomas Bishop's company, Cooper & Co. (see p. 14). They enjoyed the status that comes with responsibility for monetary transactions. Their living quarters were a small room at the back of the store. Most of them were young literate single men in their 20s and 30s.

The police records are a good source of information about the railway stores. Not only did they sell wares such as boots and underpants, tinned salmon, bottled cabbage, ham, mutton, lemonade, tobacco, jam, cake and bread from the railway bakery at Kinlochailort; they were also used by the constables for putting up public notices. When accidents occurred, the stores even provided a convenient place to shelter the injured and the dead pending the arrival of a doctor or some other figure of authority. The stores were also frequently the scenes of crime, generally theft, but also assault. In fact, all twelve shops are referred to by location in the police records: Kinlocheil, Scandale Bridge,[2] Craigag, Glenfinnan, Lechavuie, Loch Eilt, Arienskill, Kinlochailort, Polnish, Kinlochnanuagh, Glenbeasdale and Borrodale. None of these shops were selling alcohol.

Canteen staff – serving the booze

Towards the end of the 19th century, contractors began to realise that their workers' consumption of alcohol could seriously affect their work and increase the risk of accidents. According to the *Oban Times* of 1 April 1899, twenty-eight navvies employed on the Mallaig Extension had died the previous year due to 'overdosing themselves with whisky'. The fact that this 'whisky' often came in the form of raw spirits from illicit local stills, did not help the situation.

Robert jnr. had observed that 90 per cent of accidents occurred on Mondays, when the workmen were suffering from the effects of a weekend's drinking. In an effort to address the problem, he decided to open licensed canteens near the works to offer an alternative to the raw spirits sold in local shebeens.[3] The profits were put into an accident fund that was used to open a field hospital for railway workers who suffered injuries (Russell, p. 79).

'Canteen' or 'shebeen'? Location: near Polnish (HC092)

---

[2] Scandale Bridge is presumed to be Drochaid Scainnir (translates as Bridge of Scandal/Defamation). It crosses the river Dubh Lighe at Drumsallie, east of Glenfinnan.

[3] Shebeen: illicit bar where alcoholic beverages were sold without a licence.

2: The Men who Built the Mallaig Railway

Enlarged section of the picture overleaf, showing the accumulation of bottles around the door of this 'shed' with windows and shutters (HC092).

2: The Men who Built the Mallaig Railway

The railway hospital at Polnish had previously served as a school. The Art & Craft-style panels above the doors are elaborately carved, each with its own inscription: To the left: 'This School and House were built at the expense of Gertrude Emma Astley AD 1856.' To the right: 'In grateful memory of Archibald McDonald, teacher Polnish School from its formation 1856 until his death in 1894 AD'. The building has since undergone alterations but is still standing (HC226).

Hospital staff – caring for the sick and the injured

In September 1897, Robert jnr. took out a lease on a disused school building, Polnish House near Kinlochailort, and equipped it as an 8-bed hospital with a medical staff of one doctor and two nurses. There was also a domestic help. This will have allowed the nurses time to focus on their professional work, suggesting an enlightened and modern regime for the times.

Young domestic help at the gate of Polnish Hospital. Note the different uniform to the nurses opposite (HC128).

The medical staff had been contracted to care for sick and injured railway workers and those injured on railway premises. Others were refused treatment, as reflected in this correspondence between the local PC and his boss in Fort William:

> Feb 8, 1900: Telegram: 'Peter McNally got his leg broken accidentally at Kinlochailort Hotel. Railway Doctor refused attendance because not injured on Railway works. [He] is lying destitute in hut [at] Arienskill.' Reply telegram: 'Have informed Inspector of [the] Poor, Arisaig who should look after destitute cases.' (PC Sinclair, Kinlochailort)

The names of two doctors appear regularly in the police records over the course of the hospital's life: Dr Patrick between February and August 1898, and Dr Moorhead between October 1898 and December 1900.

Dr Harry Couper Patrick (1871–1942) graduated from Glasgow University in 1895 as a Bachelor of Medicine, Master of Surgery. After serving his stint in the West Highlands, he moved back to Glasgow to take up a post at the Royal Infirmary. He later moved to China and ran Shanghai General Hospital with his Australian wife until he was interned by the Japanese in 1942. Harry was the younger brother of John Patrick who was to become Consulting Surgeon at Glasgow Royal infirmary and President of St. Mungo's College.

Dr George Glossop Hamilton Moorhead (1874–1954) was from Somerset. He trained in Ireland and graduated in 1897. He was therefore a newly qualified 24-year-old when he took the job in the West Highlands.

The police records also tell us about one of the nurses who was working at Polnish throughout the railway hospital's period of operation:

> Sept 18, 1897: Belle Hawkins (27), unmarried, sick nurse, residing at Kinlochailort Hotel, was assaulted by a tramp navvy. Belle says. 'Saturday 18th I went along the public road to the New Hospital at Polnish […].' The tramp accused her of selling bad drink at the Hotel. 'He gave a blow on my breast which knocked my head against the telegraph pole by the roadside. He kicked me once. He was clean shaven and had a wild expression. I should like if he be found and prosecuted.' (PC MacKay, Kinlochailort)

We meet Belle again later, in the police records of PC Sinclair:

> July 26, 1900. Proceeded to Polnish Hospital enquiring about lost dog, property of Nurse Hawkins at the Railway Hospital […]. (PC Sinclair, Kinlochailort)

Two nurses standing in the doorway of Polnish Hospital; matron Belle Hawkins, as indicated by her big bow of authority, to the left (HC139).

Matron Belle Hawkins off duty with cat and three dogs at Polnish Hospital (HC133).

2: The Men who Built the Mallaig Railway

According to the Register of Nurses (1909), Jane Weeks Allanson Bath Hawkins (1871–1936) from Bow in Devon was the matron at Polnish Hospital between September 1897 and December 1900. In 1903, she married Dr George Moorhead, latterly of Polnish, and they moved to his native Westonzoyland in Somerset. They are both buried in the cemetery there.

Many of the injuries attended to at the hospital were caused by blasting accidents, or they involved falls from cuttings or viaducts. Crushing was also frequent injury, caused by falling rocks or toppling plant and machinery. Hypothermia was another recurring problem, as workers fell over in a drunken stupor on their way back to the hut on a Saturday night.

Blasting accidents caused the most severe injuries and the most fatalities. The following is a fairly typical incident and gives a good idea of the hazards of the work:

> February 24, 1900: [...] Statement of Timothy O'Shea (54), married [...] 'I was working a fifteen cwt [762 kg] crane in a railway cutting opposite the loading platform of Kinlochailort Station. The crane was situated on the west side of the line twenty feet from the bottom of the cutting where there is a steep slope covered with about one and a half feet of moss and when in the act of lifting a boxful of stones into the service trucks with said crane, the moss gave way under the legs causing the whole crane to collapse into the cutting. We fell down with it [...]'. (PC Sinclair, Kinlochailort).

Right: Cutting with crane lifting boulders into a wagon. Close-up of crane supports above. Location: tunnel above Camas Driseach (HC034).

Opposite: The 'walking wounded'. Bandaged patients outside Polnish Hospital (HC129).

The police constables – keeping the peace

Lochaber Archive Centre in Fort William holds the daily journals kept by the constables stationed at Glenfinnan and Kinlochailort during the Mallaig Extension construction period. Part funded by the railway company, there were two constables at both police stations at the height of activities, from May 1897 to Sept 1899 in Kinlochailort, and from Sept 1897 to May 1899 in Glenfinnan. All constables reported to Inspector Chisholm in Fort William. While the Glenfinnan police station appears to have closed in August 1900, Kinlochailort police station was staffed until March 1901.

The Glenfinnan constables covered the area from Kinlocheil Post Office to Ranochan at the east end of Loch Eilt, while the Kinlochailort constables covered the area from Ranochan to Glenbeasdale. There were also police stations in Moidart, Arisaig and Mallaig.

The police constables' duties involved not only patrolling the beat, sometimes by foot, sometimes by bike, but taking witness statements in connection with accidents, rounding up runaway soldiers, accompanying arrestees to Fort William, checking that explosives were appropriately stored, and being present at the railway worker's pay office on pay day. They also kept a tally of vagrants in the area.

The majority of the crimes reported to them were thefts of money, food, drink, clothing and boots, all of which were essential items for keeping body and soul together. The testimonies bear witness to the distress caused by such crimes, which at times may well have been triggered by even deeper distress.

Any appropriation of property from the landed estates, however insignificant, was considered with the greatest seriousness and was promptly dealt with, like here, on Christmas Day in 1898:

Police Constables nos. 14 and 17, believed to be Murchison and Sinclair, possibly at their regular meeting point, Ranochan by Loch Eilt (HC125).

December 25, 1898: 7am was informed by Donald McLeod (56), married, gardener residing at Glenaladale. 'On Saturday 24th about 4pm I was working at the Roman Catholic Chapel when accused, Alexander King (34), single, navvy, Glen Glenfinnan, wanted some evergreens as he wished to decorate some of the Navvy Huts, it being Christmas eve. I informed him to go to my house and my daughter would give him a few branches. A few hours afterwards I was informed by witness Alexander McKenzie, Shepherd, that the accused had a branch of a tree, that it was one of Col. McDonald's (my employer) favourites (Aricadian trees). I at once made enquiries and found it to be the case. [...].'

Proceeded to the Glen, found the accused in McKins Navvy Hut. Cautioned, charged and apprehended the accused who admitted the charge. Took him to Glenfinnan Hotel and conveyed accused by hired conveyance to Fort William. (PC John McKenzie, Glenfinnan)

The thief was later nicknamed 'James Puzzle' because he had been convicted at Fort William Court for cutting off the branch of a monkey puzzle tree.

Monkey puzzles (araucaria araucana) are still a feature of Glenfinnan.

Given the nature of the living and working conditions, and the fact that some of the railway workers were hiding in the West Highlands on the run from the law, it comes as no surprise that assault was almost as common as theft. Charges of malicious mischief and breach of the peace were not infrequent.

However, while assault would be investigated and charges brought, there was, on the whole, little that a couple of policemen could do to keep the peace when fiery tempers were fuelled by drink, hunger and a general feeling of misery. The constables were, it appears, required to ensure fair play rather than intervene, if at all possible:

> [R McGregor to C MacKenzie:] 'A crime had to be very serious before the police would take action [...]. Their duty was to keep order and see fair play when a row started' (Mackenzie, p. 3/3).

Thankfully, McGregor goes on to reassure us that the more level-headed railway workers would generally assist the law enforcers in doing so.

The police constables kept a census of vagrants, who would offer their labour for food and a night's shelter in an outbuilding (HC132).

The Oberon 'beached at tide' at Camas Driseach at the head of Loch Ailort. Note the horse and cart alongside (HC132).

3 The Power of Water, the Strength of Concrete

How the Mallaig Railway was built

Logistics

Building the 40-mile Mallaig Railway through inhospitable, remote countryside involved major logistical challenges. There was hardly any pre-existing infrastructure, so how was the contractor to feed and shelter 2000 men, and bring the necessary machinery to the construction sites?

The McAlpines made use of what was naturally available to them: the sea lochs. The area's one road was clearly unable to meet the company's transport requirements, so the many coastal inlets were used to access camps and construction sites with goods, plant and supplies brought in on cargo vessels.

The first 10-mile stretch out of Corpach was constructed relatively easily, along the northern shore of Loch Eil, with bridges and seawalls largely completed by July 1897. A jetty was then built at the head of the loch and the Glasgow-owned cargo ship Wharfinger, and two barges, were hired to go back and forth with goods from a works depot at Corpach.

To the west, supplies were brought by sea direct from Glasgow to camps built near Lochailort, Loch nan Uamh and the Morar estuary. Where no jetties could be built, flat-bottomed 'puffers' were useful:

> The material for the railway was brought in boats which were beached at tide and unloaded with extreme speed before the tide turned (MacKenzie, p. 3/2).

From there, a contractor's railway was built to the work sites, so that supplies could be transported to where they were needed. Where rail access was impossible, horses were used, up to 200 at a time, purchased at the so-called 'tinkers' fair' in Fort William (Hardie, p. 37).

By organising the logistics in this way, the McAlpines were able to save time by starting construction works simultaneously in multiple locations:

Section start-up months:					
Corpach	Kinlocheil	Glenfinnan	Kinlochailort	Beasdale	Borrodale & Morar
Feb 1897	Feb 1897	July 1897	March 1897	July 1897	July 1897

Source: Engineer's Certificates

FATAL ACCIDENT — It would not be unreasonable to think that the shuttle traffic on Loch Eil was relatively risk free for the boat crews, but in January 1899 PC McKenzie reported a devastating accident. While the Wharfinger lay anchored at Kinlocheil Pier, her five crew members came to grief as a result of fumes leaking from an oil lamp. They were found unconscious by the ship's captain as he came to wake them up. Attempts to resuscitate the ship's mate and chief engineer were unsuccessful.

3: The Power of Water, the Strength of Concrete

The Gordon moored by the mouth of Morar River (HC223).

3: The Power of Water, the Strength of Concrete

The Gnome moored on the northern shores of Loch nan Uamh (HC053).

Drilling

The unforgiving landscape along the route of the Mallaig Railway was not only difficult to access; it was difficult to excavate. The normal manual drilling method involved one man sitting with a drill held between his knees while two others banged it with hammers (Hardie, p.37). Here, the bedrock was predominantly mica schist, quartz and gneiss of a type that challenged even the sharpest steel and the latest tools. The McAlpines were reportedly the first contractors in Scotland to use pneumatic drills, on their Lanarkshire & Ayrshire Railway contract (1884–1890), but the rock local to Lochaber was in a different league:

> In the Highlands these [compressed-air] drills were quickly blunted [even when made of] the finest and hardest steel. Progress was slow (Hardie, p. 37).

The pneumatic drills were powered by steam-driven air compressors and transporting coal to the construction sites was costly. The new drilling technique was therefore too expensive to replace manual drilling altogether. Blasting was an alternative to some of the drilling, but gelignite was also expensive, and frequently caused accidents, such as the one that blinded John Conelly:

> 'I along with Thomas Welsh, labourer [...] and Patrick Murphy, Ganger, were boring a hole in a rock cutting at Leckavoie. I and Patrick Murphy were striking and Thomas Welsh were holding the drill. Some water was gathering about our feet. I took hold of a pick and began making a small channel to get the water away and prevent it from entering the hole we were boring. When putting the pick through the ground, there was a sudden explosion and the pick was thrown or fell out of my hands. I cannot recollect which. By the shower of sand and gravel I was rendered blind. [...] How a detonator came to be there I could not tell'. (1 February 1899, John Conelly to PC MacGruer, Glenfinnan)

Such occurrences reduced morale and aggravated the shortage of manpower. Plans for cuttings were modified and turned into plans for tunnels in order to minimise the need for blasting. By the winter of 1897/98, however, the contract was falling seriously behind schedule and the cashflow problems were severe.

Pneumatic rock drill. Note spanner and mug amongst the rubble (HC013).

3: The Power of Water, the Strength of Concrete

Part cutting, part tunnel near Lochailort (HC021).

Overleaf: Steam-driven compressor and receiver tank for storing compressed air, at the east end of Loch Eilt, where the River Mhuidhe runs into the loch (HC016).

39

3: *The Power of Water, the Strength of Concrete*

Blasting

The engineers had originally intended to get through to Mallaig with only three tunnels. In the end, there were eleven, each between 22 and 346 yards long. Plans were changed once it was clear just how hard the local rock was, but a total of 100 cuttings were still required. The explosive used was gelignite, which was considered safer than dynamite as it was less volatile and needed a detonator to set it off. Nevertheless, it caused a series of serious accidents. Outcrops of rock were removed by drilling a series of holes in them, placing a charge in each hole and then setting all the charges off at the same time. If a fuse was faulty, the charge was sometimes delayed and would go off as the men moved in to clear away the rubble. Other charges might not go off at all until an unfortunate navvy, like Thomas Cullen, struck it with his pick:

FATAL ACCIDENT – A blasting accident on 9 April 1900 cost the life of 22-year-old Thomas Cullen from Wexford in Ireland. His pick came into contact with a gelignite cartridge left behind after a blasting operation some eight months earlier. Before he passed away, Thomas explained to PC MacGruer: 'All of a sudden an explosion took place, the shock of which threw me to the north side of the cutting.' He lost both his legs and his eyesight. The following month, an inquiry was held into his demise. The verdict: accidental death.

The eleven tunnels were mostly unlined, but where necessary, 12" thick concrete lining was used (ref. Major Pringle's Line Inspection report, 1901).
Location: Twin tunnels west of Loch nan Uamh viaduct (HC022).

Filling

The rubble that was produced from making cuttings and tunnels was crushed into suitably sized aggregate by mechanised stone-breakers (see p. 117) and carried by rail to sites where it was needed, either for mixing with cement to form concrete, or for strengthening embankments. This 'cut and fill' process was standard practice as a way of minimising haulage and labour.

Rubble from cuttings being tipped to strengthen the embankment above the northern shore of Loch nan Uamh, changing the trackbed in the process from that of a temporary contractor's railway to permanent way (HC065).

3: *The Power of Water, the Strength of Concrete*

Floating

Between Arisaig and Morar the construction problems were the opposite of what had been encountered further east: the ground was not solid enough to carry a railway. To get the line across the soft and peaty stretch of land known as Keppoch Moss, the contractors used the same principle that had been used to get the West Highland Line across parts of Rannoch Moor: floating on a subsurface raft made from alternate compacted layers of turf and brushwood, capped by a large quantity of cinders.

According to the paybills, brushwood was also used for the line section at Carnach, just east of Larachmore Viaduct near Arisaig.

Peat cutting field between Morar Viaduct and Morar Station (HC067).

The power of water

Inspired by observing his Helensburgh dentist at work with a water-powered drill, young Thomas Malcolm McAlpine (see p. 14) started to investigate the possibility of using the water in the lochs along the railway line as a source of power to drive the pneumatic rock drills:

> [Thomas Malcolm] noted that instead of working the drill with the usual pedal, the dentist was merely pressing a knob on the floor. Upon inspection he discovered that the 'Pelton wheel' was driven by water; a water pipe had been cut under the floor and when the knob was depressed the water hit the wheel and drove it round. On his way home the young man thought over this new use of the water wheel principle. He saw in his mind all the lochs near the line and wondered if water could not as easily drill the rock as his teeth (Hardie, p. 37).

The company's chief engineer, Andrew Reid, felt the idea was worth trying and promptly ordered a water turbine, with a compressor coupled to it, to be sent north from Garrick & Ritchie, the water-power experts in Edinburgh. This was in the winter of 1897/98 and the contract was running seriously behind schedule – anything that might help was worth trying.

The hydropower context

It may be worth pausing the story here to look at the historic context of Thomas Malcolm's idea and Andrew Reid's decision.

The water-driven turbine was introduced around the middle of the 19th century, and in 1880 Lester Allan Pelton patented his famous wheel, which extracts energy from the impulse of moving water. In the decades that followed, hydroelectricity was adopted in numerous places in Britain for use in mills and for lighting private mansions, amongst them Kinloch Castle on Rhum. In 1896, the British Aluminium Company built a powerhouse at Foyers, by Loch Ness, thereby becoming the earliest large-scale producer of hydro-electric power in Scotland, among the earliest such developments in Europe (Tucker, 1976).

In a railway context, it is worth noting that in 1898, Norwegian engineers built a hydropower station 870 metres above sea level, with an output of 2 x 130 hp, to facilitate the excavation of the remote 5,311-metre Gravhals tunnel on the Bergen-Oslo Line. The original plan had been for manual drilling, but just like in the West Highlands, slow progress through hard gneiss necessitated generation of electric power to operate pneumatic drills (Lund, 1900).

The Norwegians reportedly came up with the idea after a study tour to the Alps and a group of fifty Italian navvies were brought in to train the local men in how to operate the new drills. However, the Italians found it difficult to cope with the hard rock and the harsh weather, so were asked to return home as soon as practicable (Heber, 1924).

The McAlpines' decision to use hydropower on the Mallaig Extension therefore forms part of a general flurry of innovative developments in electrical engineering at the time, some taking place in the close vicinity of the railway construction sites, others taking place elsewhere, in even tougher climates. What makes the Mallaig Extension hydropower scheme stand out, is therefore not necessarily the engineering ingenuity, but the nerve to trust the fledgling science and the audacity to put new thinking into real-world practice – on a very tight budget. Unlike the other hydropower schemes, the McAlpine's small-scale installations on the Mallaig Extension were of a temporary nature and many of their components could be re-used elsewhere. Indeed, were it not for the damming of Loch Dubh, there would be no trace of them today, except the concrete foundations for the turbines, now overgrown and almost unrecognisable.

Loch Dubh

The first turbine bought from Edinburgh was installed near Arnabol Viaduct. Loch Dubh, between Lochailort and Beasdale, was dammed to provide a controllable waterflow for it, raising the loch by 7ft in the process.

According to *The Engineer* magazine of 16 September 1898, the breast wall built across the outlet of Loch Dubh was fitted with screens, sluices and a scour, similar to the outlet for waterworks. While the casing of the turbine is said to have been 3ft 6 inches in diameter (just over a metre), the turbine itself is described as an 11-inch (30 cm) double discharge horizontal type passing 1300 cubic feet of water per minute. It is said to have been working at a speed of 900 rpm, outputting 100 hp.

John W.C. Haldane gives a detailed description of the Double Discharge Horizonal Turbine supplied to the McAlpines by Messrs Carrick & Ritchie in his *Railway Engineering* from 1908 (see p. 115). *The Engineer* article continues:

> The [turbine] foundations and shaft were made long enough to accommodate a dynamo of 75 horse power, which conveys electricity to motors for driving rock crushers and air compressors and for illuminating tunnels. The air is conveyed along the route of the railway in cast iron flanged or malleable iron screwed pipes, the greatest distance being about nine miles at present. On the iron pipe a tee with a valve is placed where required, and into this a flexible pipe is fitted, which connects with the rock drills in the cuttings (*The Engineer*, 16 Sept 1898, p. 275).

Loch Dubh, looking west towards its outlet, trackbed on the left (HC062).

Loch Dubh, looking east towards its inlet, trackbed on the right (HC063).

3: The Power of Water, the Strength of Concrete

Above: Arnabol Viaduct with turbine house to the right, fed by water in 21" steel riveted pipes. The skeletal building in the centre is assumed to be a sawmill in the process of being dismantled, with the no-longer-needed timber arch formwork on the ground next to it (HC091). See p. 116 for comparison.

Opposite: Filling in the east end of Lochan Deabtha between Loch Dubh and Arnabol Viaduct, to allow one of the last sections of track to be laid. The path at the rear is the footpath to Peanmeanach. Note the gap in the terrain at the far end of the pool, ready for the footbridge across the line to be put in place. Note also the dam wall and sluice in the foreground (HC060).

3: The Power of Water, the Strength of Concrete

Left: 'The air [from the hydro-powered compressor to the pneumatic drills] is conveyed along the route of the railway in cast iron flanged or malleable iron screwed pipes'. (*The Engineer*, 16 September 1898). Location: Third tunnel east of Lochailort, before Polnish. Note the flanged pipes that run alongside the track (HC033).

Right: 'On the iron pipe a tee with a valve is placed where required, and into this a flexible pipe is fitted, which connects with the rock drills in the cuttings' (*The Engineer*, 16 September 1898). Note the flimsy-looking wooden trestles that carry the flanged pipes across the rough terrain, the flexible pipe connecting to the drill, and the chain from the crane that is lifting the apparatus into position (HC014).

Switching to hydropower had a major beneficial effect on the contract. The new compressors were found to be four times as efficient as the steam-driven ones. The McAlpines therefore adopted the same power-generating method at other locations, forming artificial pools wherever necessary. In 1898, they had sixty pneumatic rock drills operating with air from electrically powered air compressors. The use of hydropower reduced the need for manpower by 500-600 men, saving both the budget and the schedule in the process.

Above and overleaf: Turbine house at the foot of the Mhuidhe falls (HC057).

Lechavuie

An artificial pool was built at Lechavuie, near the entrance to the twin tunnels west of Glenfinnan, where the river Mhuidhe starts cascading down towards Loch Eilt. The dam was adjacent to the site of a navvy hut and one of Cooper & Co.'s stores. The turbine house was sited at the foot of the falls.

Most navvies will have been unfamiliar with turbine wheels, and this unfamiliarity with the new technology caused accidents:

> February 14, 1900: James Kelly (21), labourer [...] 'I was accidentally injured the previous night by getting entangled in a turbine wheel used in connection with the air Compressor at Lechavuie. I did as I was bid, putting my foot on the wheel. I brought the high side down so that the wheel was lying level. Mathew Blair went outside and immediately after doing so the wheel started and my left foot slipping into it caused me to be dragged round for a few yards after which I was pitched clear of the wheel which was very fortunate for me. This was my first night working at the Compressor'. [...]. (PC McGruer, Glenfinnan)

Another accident with more serious consequences took place above the waterfall, by the dam. It was however caused by Mother Nature, whose forces proved too fierce for the ramshackle buildings of the railway contractor:

> FATAL ACCIDENT — In the early hours of 11 March 1899 the dam at Lechavuie burst, and the water swept away the Cooper & Co store and all its content, including its young shopkeeper, 25-year-old Alexander Mactavish, a promising athlete from Rothesay. He had been sleeping in his room at the back of the store when disaster struck. His body was recovered from a lower reservoir which had impeded the torrent from the higher dam. The accident was reported by the local newspapers to have been caused by the sudden melting of snow.

The strength of concrete

The hardness of the local rock (mica schist, quartz and gneiss), was the reason why engineers Simpson & Wilson had decided on mass concrete[1] as a building material for the engineering structures on the Mallaig Extension before the construction contract was awarded. Walter Stuart Wilson makes this clear in his paper to the Institution of Civil Engineers, entitled 'Some Concrete Viaducts on the West Highland Railway':

> [...] while the stone was admirably suited for concrete, it was quite impossible to use it for masonry on a large scale. Even supposing that the requisite number of men accustomed to work such stone could have been secured, the cost would have been prohibitive.

The Scottish engineer who had done more than anyone to promote mass concrete as a suitable and cheap building material for railway structures, was John Strain (1845–1931), who had supervised the building of the Callander & Oban Railway between Dalmally and Oban and worked as Engineer-in-Chief for the Lanarkshire & Ayrshire Railway between 1888 and 1891. He had made extensive use of concrete as a building material in both contracts. Robert McAlpine had been the chosen contractor for the latter and no doubt benefited from John Strain's guidance. Sadly, problems relating to the contract caused the two men to fall out (Hardie p. 38).

At the time the Mallaig Railway contract was awarded in 1897, most railway engineers were still worried about the strength of concrete. Robert McAlpine had been using the material for housebuilding since the 1870s, when Portland cement had started to become an increasingly standardised product. In 1892–94 the company had also been using concrete to line the tunnels of the Glasgow underground between Buchanan Street and St George's Cross, which had impressed the firm of engineers in charge of that contract: Simpson & Wilson. This undoubtedly stood the contractors in good stead when bidding for the Mallaig Railway contract.

According to Wilson, the concrete they used had a ratio of cement to aggregate of 5:1 for foundations, 4:1 with displacers[2] for abutments and piers, and 4:1 for arches. No imported sand was used in any of the concrete, only the on-site stone crushed by mechanised stone-breakers to give the desired size of aggregate (see p. 117). At Glenfinnan Viaduct, which unlike the other viaducts was founded on both rock and clay, two ½-inch steel plates were inserted at the crown of each arch to allow for expansion and contraction.

While building in mass concrete may have made perfect commercial sense, many of the landowners whose land would be traversed by the new railway were less than impressed with its appearance. Concrete was considered to be unforgivably ugly, and the contractor was forced to go to great lengths to make it prettier, by adding red colour to the mix, and by scoring the surface, to emulate the look of dressed granite from a distance.

The owner of Arisaig Estate decreed that any viaduct that could be seen from his house would need to have granite piers, abutments and parapets, 'to preserve the grand appearance of the glen'. This proviso presented the engineers with a problem at Borrodale Burn.

---

[1] Mass concrete is poured or cast-in-place concrete with no steel reinforcement but large amounts of crushed stone aggregate. The material had been used for many years to build foundations for buildings and bridges and in the construction of docks and harbours, but railway engineers were slow to realise its potential in railway construction (Russell, p. 49).

[2] 'Displacers', also referred to as 'plum', are large stones or boulders used as filler material, introduced in the interest of economy, to avoid the use of excessive amounts of cement and costly small-size aggregate.

3: The Power of Water, the Strength of Concrete

Borrodale Viaduct

The engineers had originally designed a viaduct with 86-feet high piers to take the line across Borrodale Burn (Russel, p. 81). When the landowner insisted on expensive granite facing to piers and abutments, the estimated building cost sent the budget into a realm beyond the realm of financial possibility. The McAlpines were therefore asked to come up with a cost-saving solution.

They found a daring one: they proposed bridging the glen in just three spans; two short ones to the side and a wide central one of 127 feet 6 inches (38.8 metres). This would keep the number of granite blocks and masons required to a minimum. It would also mean building a longer mass concrete span than had ever been attempted before for a railway bridge.

Simpson & Wilson gave the proposal their go-ahead but insisted that the risk would have to be carried by the contractor. Work started in July 1898 and was successfully completed by November of that year, with the parapets finished off later, in June 1899. Alexander Simpson was officially credited with its design (Tyrell 1909, p. 94).

Robert McGregor was mightily proud of the achievement on McAlpine's behalf. This is how he explained the work to Compton MacKenzie:

> [McAlpine] had always grudged the expense of the heavy timber framework used to support a bridge while under construction and afterwards destroyed. So now he merely built a light framework that was used to support a thin layer of concrete which was then laid down and allowed to set with a rough surface. When the first layer had hardened, a second layer was applied with a rough surface and this process was repeated until the thickness required had been reached. The engineers were still sceptical and demanded that a hole should be cut through the bridge. The result was a triumph for McAlpine's theory. No trace of the layers were perceptible. The wet concrete had seeped into the rough hardened concrete of every layer and thus a solid mass had been formed (MacKenzie, p. 3/7-8).

Walter Stuart Wilson explained much the same procedure in his article 'Some Concrete Viaducts on the West Highland Railway', although adding some numeric precision and leaving out the contractor adulation:

> The large span springs from the solid rock on each side, and the extreme height is 80 feet. The arch, which is 4 feet 6 inches thick, is composed of 4-to-1 concrete, which was put on in layers, the top of each layer being left as rough as possible. After completion, a test-hole was cut in the arch to prove the solidity of the work; no trace could be seen of the joining of the different layers, and the concrete was perfectly homogeneous. In this viaduct no joints were made for expansion and contraction; and the result has proved that in this case nothing of the kind was required. The arch shows no indication of movement of any kind, and is absolutely without a crack. The parapets and side spans were built of stone to satisfy the wishes of the proprietor of Arisaig Estate. The total sum paid to the contractor for this viaduct was £2,109 (Wilson, 1907).

It is ironic that this significant engineering achievement came about because a landowner could not bear the sight of a grey concrete structure from the windows of his parlour. For today, Borrodale Viaduct is all but invisible due to the vegetation that has grown up around it. Despite its place in engineering history, no crowds come flocking to see this majestic bridge, like they are on the hills surrounding the much-celebrated all-concrete viaduct at Glenfinnan.

Opposite: Borrodale Bridge built July – November 1898 with a record-breaking concrete span of 127 feet, 6 inches. Granite piers, parapets and abutments (HC048).

3: The Power of Water, the Strength of Concrete

Glenfinnan Viaduct – a monument to concrete

Excavations for Glenfinnan Viaduct started in July 1897, with work on the piers beginning in September of that year. In October, the staging started to go up, with hundreds of joiners employed to construct the timber shuttering and formwork. There were mechanical stone crushers (see p. 117) working at each end, and the resulting aggregate was mixed with the cement. By October 1898 the work had progressed far enough for a contractor's railway to be laid across the viaduct. By the time the structure was completed, a total of 14,914 cubic yards of concrete had been used in its construction. The contractor was paid a total of £18,904 for building the viaduct, of which £17,883 was payment for concrete.

The design details for Glenfinnan Viaduct are well documented:

Length	Max. height	Arch spans	Pier thickness, top	Curve radius
1248 ft	100 ft	21 x 50 ft	18 x 6 ft + 2 x 15 ft	12 chains
380 m	30 m	21 x 15 m	18 x 1.8 m + 2 x 4.5 m	241 m

In 1901, Glenfinnan Viaduct was the longest concrete bridge in the UK. For all the engineering prowess and gruelling labour that went into its construction, it is however its evocative beauty and elegance that has made it a lasting monument to concrete. The slender pillars and curved line at the head of Loch Shiel, directly opposite James Graham's monument to the Jacobite rebels, provide a truly breath-taking view.

When Lady Judy McAlpine unveiled the Transport Trust's Red Wheel plaque commemorating Glenfinnan Viaduct in August 2019, she referred to it as 'the unsurpassed jewel in the crown' of Robert McAlpine & Sons' achievements.

Glenfinnan Viaduct built 1897 – 1898, opened 1 April 1901 (HC108).

The rebels of Glenfinnan

Robert MacGregor was convinced that the success of the Mallaig Railway contract owed much to the fact that the name of the firm was McAlpine, 'an ancient and famous Highland name'. To his mind, local people therefore saw the company as a local enterprise.

Whether or not that was the case, Robert McAlpine was clearly fascinated with the Jacobite cause of the House of Stuart and he took great pride in his family tartan. Like many, he saw Prince Charles Edward Stuart, who rallied the clans for the Jacobite rising in 1745, as a Romantic hero. But while 'Bonnie Prince Charlie' was a Roman Catholic, like large parts of the population in the areas west of Loch Eil, Robert McAlpine and his family were devout Presbyterians and they are said to have felt deep allegiance to Queen Victoria from the House of Hanover.

FATAL ACCIDENT – On Christmas Eve of 1898, 20-year-old Walter Crowe from Limerick in Ireland suffered a gruesome accident at Glenfinnan. He slipped while working the slope of a cutting with an iron crowbar. As he fell 25ft (7.6 m) to the bottom of the cutting, the crowbar 'entered the lower part of his body and emerged at the shoulder blade'. The bar was removed on site and the injured man taken to the Belford Hospital in Fort William.

Poor Walter died on 16 January 1899 after suffering weeks of excruciating pain. (Sources: PC MacGruer and the *Dundee Courier*)

James Gillespie Graham's monument to the Jacobite rebels who gathered in Glenfinnan in 1745 to follow Prince Charles Edward Stuart on his military campaign to reclaim the British throne for the Stuarts. The monument was erected by Alexander MacDonald of Glenaladale in 1815 (HC059).

3: The Power of Water, the Strength of Concrete

Loch nan Uamh Viaduct: the horse & cart in the pier

Excavations for the Loch nan Uamh Viaduct started in June 1897, with work on foundations and piers beginning the same month. Staging and formwork were put up in October 1897, the same month as for Glenfinnan Viaduct. When construction was completed in October 1899, a total of 4,557 cubic yards of concrete had been used. However, according to the Engineer's Certificates, work on the viaduct appears to have stopped completely between October 1898 and August 1899.

The Engineer magazine of September 1898 reported that the 8-arch viaduct at Loch nan Uahm is 'similar in construction to that over the Arnabol Burn'. It is however evident that the 54ft-wide middle pillar of the viaduct at Loch nan Uamh makes for a rather different design than the 6-arch viaduct at Arnabol (see p. 47), and it is tempting to speculate why this may be the case.

Alexander Simpson's great-grandson James Shipway (1926–2013), who wrote extensively on Scotland's civil engineering heritage, pondered as follows in 1998 when he gave a talk on the construction of the West Highland Line to the Institution of Engineers and Shipbuilders in Scotland:

> [...] Surmise suggests that originally only a 4-span viaduct was planned over the public road and the river, with the bridge abutting an embankment on the rather higher ground to the east. It could have been that this prominent embankment was the subject of second thoughts after the east abutment to the 4-span western part of the bridge was already constructed, for such an embankment would have been an unwelcome intrusion and considerably restricted the view. This could have led to the substitution of the 4 eastern arches for symmetry, the abutment already constructed being left in place, since its demolition might have endangered the stability of the original 4 western arches. Such a surmise however must be recognised as conjecture (Shipway, 1998).

Loch nan Uamh Viaduct, built June 1897–October 1899 (HC052).

What we do know, as evidenced by pictures held by the National Railway Museum (see p. 118), is that the four arches to the west were completed, and the formwork removed, while work was still ongoing on the four arches to the east.

What we also know, as evidenced by the radar imaging examinations conducted by the Institution of Civil Engineers in 2001, is that the middle pier contains the remains of a horse above the wreckage of a cart. A heavily laden cart must have toppled into the pillar with its rubble content and pulled the animal with it. Local legend had long had it that a horse had fallen into Glenfinnan Viaduct, but the radar imaging exercise proved that Loch nan Uamh was in fact the scene of the accident.

In a conversation with John Barnes of Glenfinnan Station Museum after the remains of the horse and cart had been discovered, James Shipway mentioned a new alternative theory. He wondered if perhaps the horse-and-cart accident happened in the panic that arose after the collapse of a previously-existing arch. If so, the original design would have been for eight even 50-ft arches, like the Arnabol Viaduct's six, with no 54-ft middle pier and without the current easternmost arch. As Shipway pointed out, the low arch to the east barely clears the rock outcrop, which would instead have formed a suitable abutment.

If this alternative conjecture is right, then what appears on record to have been an idle year, might have been used to redesign and rebuild the viaduct at Loch nan Uamh. We will probably never know what actually happened. The Engineer's Certificates record only the progress made, not the setbacks or the circumstances. To our knowledge, there is documentary evidence only that:

- in September 1898 the viaduct was reported to be 'of similar design to the one at Arnabol',
- work halted between October 1898 and August 1899,
- a total of 517 cubic yards of excavations were carried out for the viaduct, of which 189 were carried out after the pause, and that
- the middle pier holds the remains of a horse and cart.

Driver with horse and cart by Lochan Deabtha (HC038).

Mallaig pier – part-funded by Parliament

Although the Houses of Parliament at Westminster originally approved the building of the West Highland Line in order to alleviate the poverty of the coastal population, it proved difficult for the promoters to find sufficient capital to take the railway all the way to the western seaboard. In 1896, after two frustrated attempts, Parliament therefore stepped in and granted shareholders a guaranteed annual dividend of 3 per cent on a large portion of the capital, for a period of thirty years from 1901. This measure soon had the desired effect of stimulating share subscription.

To make sure that the new railway would have the intended impact, Parliament also granted a sum of £30,000 towards the cost of constructing a pier and breakwater at Mallaig, two thirds of the estimated total. The sum would be payable on completion of its construction.

Mallaig Pier was the last section of the railway to see work commence. According to the Engineer's Certificates, the construction period was from April 1899 to February 1901.

Unlike the other concrete engineering structures on the line, the pier at Mallaig was built with precast concrete blocks rather than layers of poured concrete. The blocks were however cast on site, utilising machinery stationed at the water's edge.

Left:
Mallaig Harbour, looking north.
On-site casting of concrete blocks
(HC015).

Opposite:
The building of Mallaig pier using
pre-cast blocks of concrete (HC025).

3: The Power of Water, the Strength of Concrete

Divers in Mallaig harbour getting ready to assist with positioning the concrete blocks (HC011).

3: The Power of Water, the Strength of Concrete

Fully submerged concrete block held by the crane. Diver's boat to the far left, providing air by manual pump to the diver under water (HC029).

Public relations

The McAlpine family's biographer, Compton MacKenzie, believed that the comparative absence of hostilities during the construction period was down to the fact that so many of the Highlanders and Islanders were of the same faith and sentiment as the itinerant Irishmen, and he records Robert MacGregor as saying:

> The Irishmen accepted [the Highlanders] as men of the same race. All were for Prince Charlie and the Jacobite cause. [...] That was probably the strange link between the Irishmen and the Highlanders. The Irishmen respected them, too, on account of their rebel traditions, and they settled down with them well (MacKenzie, p. 3/4).

Still, there were many rows and many grievances.

Fences

In the autumn of 1900, Lieutenant Hugh Blackburn at Annat Farm in Corpach raised a complaint on behalf of himself and other sheep farmers and crofters along Locheilside. They were unhappy about the wire fencing that had been put up to stop their livestock trespassing on the railway. The space between wires was considered to be too wide.

When their first complaint was rebuffed by the West Highland Railway Company, Lieutenant Blackburn wrote to his lawyers:

> I understand that on the representation of Mrs. Head of Inverailort, the fence westward from Leac a Mhui has been spaced differently, and that no sheep have been killed on her land [...] I would also state that I have seen a large Ayrshire ox of mine with abnormally big horns walk through the railway fence without apparently being aware it was there. [...] [The company's] happy-go-lucky statement that they 'think there will be no trouble' when the line is working regularly, won't do – it is the business of the Company to see that there can be no trouble; and anyone but a Railway Engineer knows that the only difference will be that [...] the surfacemen will get their mutton gratis, this no doubt saving trouble to the Company (National Archives, MT6/1034/2).

Despite the lieutenant's high standing being pointed out in the lawyer's covering letter to the Board of Trade, the farmers' complaint was not acted on. The Government Inspector, Major Pringle, took great care to report on the fencing, but he considered the design to be perfectly adequate and in keeping with the fences used on the main part of the West Highland Line.

Level crossings

The new level crossings posed another problem for the local farming population. In March 1901 Reverend Crawford at Kilmallie Manse wrote to the area's MP on behalf of his neighbour, the local mill owner:

> [...] Our mill at Annat is situated about ¼ of a mile from the public road. The Mallaig Railway runs parallel to and close to the public road, crossing the road from the public highway to the mill. This crossing is effected not by bridge but by level crossing with the usual two gates. These two gates must be kept locked. How are the people to get their carts to the mill? There is no way of signalling to the mill when the key is required. A man cannot leave his horse and cart on the road and walk down to the mill for the key.
>
> This is a case of a Railway Co. blocking a public road – a road largely used by all the farming people in Lochaber, this being the only meal mill within many miles. It is utterly unjust to the farming interest and to the miller to hinder access to an institution so indispensable as a mill of this kind [...] (National Archives, MT6/1034/2).

This problem was communicated to the Board of Trade's Inspector, who solved the problem by giving keys for the gate to frequent customers of the mill. No documents suggest what customers with less frequent business at the mill should do.

Right: Improvements to roadside fencing from Beasdale bridge towards Arisaig House, 'for the protection of the public' (HC078).

In February 1900 Robert McAlpine was approached by the County Council and asked to make amends for 'excessive traffic placed on the road between Banavie and Morar Bridge' (Oban Times, 3 February 1900).

Left: Granite facing of Borrodale Bridge as requested by the landowner (HC024).

Peace-brokering

The McAlpines themselves were keen to broker peace, and they used the resources available to them to get the local communities on side.

When Queen Victoria died in 1901, the date of her funeral, Saturday 2 February, was declared a national day of mourning. Memorial services were held throughout the country. The service local to Lochaber was held in Corpach, at Kilmallie Parish Church. To enable the villagers of Glenfinnan to attend, the McAlpines put on a special train on the soon-to-be opened railway:

> [...] There was a large congregation, some of whom had driven long distances over roads rendered somewhat heavy by the continuance of the snowstorm. Through the kindness of Messrs Robert McAlpine & Sons, railway contractors, the people of Glenfinnan were conveyed by train to and from the service – another of the many kindnesses received by our community from the members of this firm. [...] (*Oban Times*, 9 Feb 1901)

The members of the McAlpine family were clearly doing their best to keep the lineside communities happy, and when Robert jnr. and his family left Corpach in September 1900, their departure was reportedly met with 'great regret in the local community' (*Oban Times,* 15 Sept 1900).

They must however have been acutely aware that the local population was keen to see the railway finished. Reports about good progress and hopes for a speedy completion appeared regularly in the press from 3 February 1900 onwards.

Almost there ….

Despite the McAlpines' apparent optimism, Simpson & Wilson reported slow progress to the Directors of the West Highland Railway Company. On 6 March they feared the line would not be ready for opening until October.

The McAlpines on the other hand, must have put all their faith in completion by June 1900. They even terminated their lease on Polnish House with this in mind, and had to move their medical facility to Kinsaddle near Morar. The hospital continued its work from there until the end of the year[3].

The McAlpines even organised a celebratory train to run in June, inviting family, friends and associates. The *Oban Times* reports:

> On Saturday last the first through train was run from Banavie to Mallaig. A pioneer train gaily bedecked with bunting, having a saloon carriage attached, completed the journey of 40 miles in a little over two hours, notwithstanding that portions of the way are still unfurnished with permanent rail.
>
> The pioneer train, which was occupied by members of the contractor's firm, Messrs Robert McAlpine & Sons, by engineers, railway officials, and private friends, was hailed with demonstrations of welcome by the populations along the route, and the opening of the railway for public traffic in October is being looked forward to with much expectation. [...]. (*Oban Times*, 16 Feb 1900)

---

[3] Surmised based on nurse Hawkin's entry in the Register of Nurses.

Opposite: Mallaig's West Highland Hotel under construction. Note the back-end of the moving dog to the bottom left (HC086).

In fact, progress was thwarted in the second half of 1900 for a number of reasons. The engineers repeatedly refer to a shortage of labour, and there were significant problems with the signalling installations due to inclement weather. An application in August from the Board of Directors for approval of the company's signalling plans for Banavie Swing Bridge, was rejected by the Board of Trade on grounds of safety, thereby putting back the anticipated opening day even further.

Two months later, it was an impatient Board of Trade that approached the company about a likely opening date. The hesitant response is dated 23 November:

> We hesitate to fix a date for the probable completion of the Line. The signal people experience great difficulty in doing their work at this season, and the progress in the dwelling houses for the men is slow. We should say that, at the earliest, the line cannot be ready before the end of January (National Archives, MT6/1034/2).

As we know, it took a little longer still. On 11 March, Inspector Pringle and his band of assistants eventually started their painstaking work to scrutinise the Mallaig Railway and its state of readiness for public travel.

Major Pringle's inspection

On 12 March, *The Scotsman* reported that the Board of Trade's inspection of the Mallaig Railway had begun:

> […]. The Board's representative, Major Pringle, was accompanied by Mr. Deuchars, North British Railway Superintendent, Mr Holmes, locomotive superintendent, Mr Bell, engineer-in-chief; Mr Clements, telegraph superintendent; Mr George Innes, district superintendent; and others. The bridges and permanent way were subjected to the most stringent tests by having run over them with several heavy locomotives coupled together, while the signalling and telegraph apparatus were minutely inspected. […] No definite date has been fixed for the opening of the new route, although the company hopes to have trains running by the end of this month. (*The Scotsman*, 12 March 1900)

The work took the inspectors three days to complete. Their detailed report, held in the National Archives at Kew, pays specific attention to the reasons why such extensive use had been made of 'Portland cement concrete', and makes a point of highlighting that at the time of testing, the concrete had been allowed to settle for more than 18 months. Special mention is also made of Borrodale Viaduct as being 'the largest concrete arch built in the United Kingdom' and that it was therefore rigorously tested by running three coupled goods engines over it at about 30 miles an hour. This is said to have produced only 'the very slightest vibration'.

On 18 March, Major Pringle reported to the Board of Trade that he was happy for the line to be opened for traffic, provided seven items listed for improvement were dealt with. Four of these were signalling-related, the other three concerned appropriate branding of station lamps, the fixing of some loose wire fencing, and the introduction of latch fastenings for public level crossing gates.

Permission to proceed

The North British Railway Co. ran the first service train through to Mallaig on Monday 1 April 1901, to very little cheer. The celebrations had already been taken care of the previous summer.

However, the *Oban Times* carried a piece on 6 April which philosophises at length on the event, referring to the railway's brutal imposition on a territory 'that ardent spirits hold sacred'. The article is dripping with Romantic emotion:

> […] Yet even the brutal railway may act as a stimulus to sentiment! It will bring into the midst of these treasured scenes thousands of pilgrims, who will not turn their eyes upon Glenfinnan and Loch Shiel and Borrodale and Keppoch without catching something of the spirit of the past, without appreciating deeply the splendid devotion of the clans to Prince Charlie. (*Oban Times*, 6 Apr 1901)

While today's visitors may not recognise themselves as 'pilgrims' and may indeed know nothing of Prince Charlie, the beauty of the landscape they travel through on the Mallaig Railway clearly continues to impress.

Ultimately, it was the fact that the railway offered an opportunity for the area's inhabitants that allowed the *Oban Times* reporter to reach his positive conclusion:

> [...] A fresh and speedy avenue to the great markets of the south puts a new weapon into the hands of the inhabitants which they may use for their own advancement. [...] Before long, it must come to play a large part in the life and work of the Highlands. (*Oban Times*, 6 Apr, 1901)

The author of this celebratory article in the local newspaper was of course proven right: although the Mallaig Extension to the West Highland Line did not fulfil all expectations, it did bring about a positive change in the lives of the area's population, and luckily the line continues to play a significant role in the Lochaber economy today.

It took ingenuity and skill, persistence and audacity, hard toil and human suffering to build the Mallaig Railway – as indeed has been the case with most of the world's great railways. It also took political manoeuvring and considerable funds, public as well as private.

The future

The Mallaig Extension is a magnificent piece of infrastructure gifted to us by a past generation. In the face of today's environmental challenges, we need to match the far-sightedness of those who came before us, so that we optimise the line's usefulness for Lochaber's current population. But when enhancing the line for future use, it is essential that we also grant it the long-term protection it needs as a cultural heritage site.

At the very beginning of the 20th century, the Mallaig Railway demonstrated to the world the benefits of concrete as a building material. Reborn from its Roman past thanks to the invention and gradual standardisation of Portland cement, concrete proved itself as a material suitable for use on a large scale. Thanks to the dogged resolve and aesthetic sensibilities of its engineers and contractors, the Mallaig Railway bears witness to the strength, affordability and beauty of concrete-built structures and their capacity to enhance and complement a scenic landscape.

Overleaf: Arnabol Viaduct from the path to Peanmeanach (HC064).

4 The Person behind the Camera

A photograph is a point of view

The photographs in the Holden Collection have given us new knowledge about the people who built the Mallaig Railway and how they went about building it. But a photograph will always represent a point of view that reflects the photographer's outlook. It is important, therefore, to know whose perspective the pictures invite us to share. Because we allow this photographer's work to inform us about a piece of history, we ought to know what may have motivated the pointing of the lens.

The Cornwall auctioneer who sold the cellulose nitrate negatives to Michael Holden could not provide any information about the vendor, and the numbering of the pictures merely reflects their sequence as bought. They are stand-alone negatives inserted into two custom-made negative albums, none of them attached to any of the others. It is clear that they are taken with the same camera, but their edges are so delicate that it is impossible to positively identify a sequence from the way the film strips have been cut.

Consequently, it soon became apparent that the only clues to the photographer's identity were in the detail of the images themselves. To aid our search, we formulated four questions:

i. When were the photos taken?
ii. Who might have had sufficient time and money on their hands to buy a state-of-the-art camera and to use it liberally?
iii. Who would have had access to all the depicted locations?
iv. Who would have chosen the subjects featured?

We addressed the four questions by closely examining the pictures and diligently researching their context. For were we to find the answers, we might at least be able to make an educated guess at the identity of the photographer.

So, we selected our best magnifying glass, invoked our 'inner Miss Marple' and set about some exciting detective work that was to take us further afield than we had ever expected. What follows is the story of that journey, with all its twists and turns.

Dating the photographs

While much of the contractor's paperwork for the period has not survived, the black leather ledger containing the pay bills, i.e. the Engineer's Certificates, is held by the University of Glasgow Archive Services. This gives us detailed information about the sequence of construction events. By looking at the state of completion of the engineering works in the photographs and comparing this to the work progress reflected in the pay bills, we found that our initial dating of 'sometime between 1899 and 1901', could be narrowed down to 'sometime between June 1899 and October 1900', at least with respect to the 177 pictures from the areas around the Mallaig Railway.

The weather

While snow on the ground and a lack of foliage suggest that the photos were taken in the winter months (November 1899 – March 1900), the ice-covered lochs in some of the pictures provide an even better clue. Temperatures rarely fall low enough for long enough for the local watercourses to freeze over. By going through the observations issued by the Meteorological Office for the winter in question, we found that February 1900 was particularly cold in Scotland, with the lowest temperatures experienced between the 8th and the 15th.

Looking at the Oban Times for 17th February 1900, we soon found that the deep freeze had been newsworthy:

SNOWSTORM – The severest snowstorm experienced in this district during the present winter took place towards the end of last week, and continued with occasional breaks till Tuesday. […] On Tuesday a thaw set in, and with a southerly wind the snow gradually disappeared […]. (*Oban Times,* 17 Feb 1900)

SKATING – Devotees of this favourite pastime have had ample opportunity during the past few days of enjoying their winter sport. The usual ponds were discarded for the superior ice on the River Nevis – which was frozen over – while a few betook themselves to the miniature lakes formed in the Blar Mhor. The Caledonian Canal and parts of the River Lochy were also frozen. (*Oban Times,* 17 Feb 1900)

Railway hospital staff having fun on the ice on Lochan Deabtha at Polnish. Skating, like curling, was a popular pastime (HC155).

The building of Glenfinnan Station

Glenfinnan Station has been home to our museum, and ourselves, for nearly thirty years. Yet it took us quite a while to realise that picture no. HC031 was in fact a picture of our own back yard: the woodland area below the station. The complete absence of trees had us thoroughly confused. But when the Eureka moment eventually came, there was no longer any doubt: the concrete underpass under the line, to the east of the station, still bears the same markings in the concrete. This is unmistakably Bridge 66, the sheep creep that links the rough pastures either side of the line at Glenfinnan.

Bridge 66 is now the starting point for the trail that takes walkers from Glenfinnan Station to Glenfinnan Viaduct (Bridge 65 to railway engineers). The square concrete structure in the background is the platform for the station's goods siding.

However, there is no sign of the station buildings.

The Engineer's Certificates tell us that the concrete walls for the loading bank at Glenfinnan were largely completed by the end of February 1900. Work on the station building started after mid-March 1900.

So, here was the verifiable proof of date we had been looking for, enabling us to say without a doubt that:

(i) THE PHOTOS OF THE MALLAIG RAILWAY WERE TAKEN IN FEBRUARY/ MARCH 1900.

Works train carrying navvies and machinery across Bridge 66, the sheep creep by Glenfinnan Station (HC031).

4: The Person behind the Camera

4: The Person behind the Camera

The camera, the time and the money

The quality of the photographs in the Holden Collection is extraordinary, allowing us to inspect the tiniest of detail. While many of them have the character of present-day holiday snaps, others are attempts at portraiture and some at reportage. The focus is often on the subject rather than on the composition, suggesting a non-professional photographer who enjoys experimenting with his gadget and its potential. The photographic equipment on the other hand, is clearly top-of-the-range and state-of-the-art. It must have been extremely costly and would have been well outwith the budget of the ordinary man or woman.

The photographer would also have to be able to afford the time to travel round the countryside and capture scenes of personal interest, scenic beauty and engineering significance on film.

This led us to hypothesise that:

> (ii) THE PHOTOGRAPHER IS FROM THE HIGHER ECHELONS OF SOCIETY AND TAKES AN INTEREST IN ENGINEERING

Access to locations

Out of the 244 photographs, 177 featured locations in the vicinity of the Mallaig Extension. As we have seen, there were photographs of engineering structures like bridges and viaducts, contractor's locomotives and plant, equipment and tools, tunnels and cuttings, men at work — motifs that would interest a railway engineer. Many of these were taken at locations that would have been inaccessible to members of the general public. There were 64 pictures of this type. While several site visits were made by railway officials over the construction period, we have found no records of any such formal visits undertaken in February/ March 1900.

There were also pictures of buildings, 23 in all, ranging from temporary wooden railway structures to churches, hotels and such like. A further 45 feature Polnish Railway Hospital and its immediate vicinity. This is the sort of motif that would interest someone who has spent time in or around the buildings depicted.

Interestingly, picture nos. HC088 and HC089 (see p. 11) are close-ups of the bungalow in Corpach where Robert McAlpine jnr. lived with his family throughout the construction period, until September 1900. On p. 13 we also saw that the photographer had access to an album with a photograph of William McAlpine, Robert snr.'s second son, who was in charge of the company's head office in Glasgow. Our photographer appears to have had a special relationship, not only with the railway works, but with the McAlpine family.

Many of the Mallaig Railway photos are panoramic vistas taken from hilltops and mountain tracks. While it was customary for people to walk vast distances at the time, someone as wealthy as our photographer would most likely have moved around in this sort of rugged terrain on horseback. Nevertheless, getting to the various photo locations will have required a level of agility and energy which suggests a youthful individual.

The range of locations featured led us to infer that the photographer is a young person who enjoys unhindered access to the railway works and is on good terms with the McAlpine family :

> (iii) THE PHOTOGRAPHER IS A YOUTHFUL AND ENERGETIC INDIVIDUAL WITH A PERSONAL CONNECTION TO THE MALLAIG RAILWAY WORKS AND THE MCALPINE FAMILY

4: The Person behind the Camera

Loch nan Uamh Viaduct: trackside location inaccessible to members of the general public (HC009).

Forty five pictures feature Polnish Railway Hospital or its immediate surroundings (HC120).

4: The Person behind the Camera

Loch Beag and Loch nan Uamh from Telford's Parliamentary Road, with a glimpse of Arnabol Viaduct (HC071).

4: The Person behind the Camera

Choice of subjects

People

We saw in chapter two that one of the most intriguing characteristics of the Holden Collection is the many portraits of people involved with the construction of the line. We meet engineers and workers, patients and nurses, policemen, tramps and even a domestic maid or two. And the demeanour of many of the people looking back at us from these photographs, whatever their social standing, is neither stilted nor formal or posed, but relaxed – some are even smiling.

Buildings, scenic beauty, railway works and pets at play

We have also seen that the photographer documents buildings, scenic beauty, concrete structures, railway works and pets at play. Whoever our wealthy young horse-riding photographer is, it appears to be someone with a wide range of interests and a real zest for recording domestic life around Polnish Hospital. There are no similar scenes that might document life in the navvy camps, so the pictures from Polnish Hospital would appear to reflect a personal rather than a social interest.

In combination, the choice of subjects featured led us to infer that:

> (iv) THE PHOTOGRAPHER, WHETHER MALE OR FEMALE, RELATES TO PEOPLE FROM MANY DIFFERENT WALKS OF LIFE AND IS KEEN TO RECORD DOMESTIC LIFE AROUND POLNISH HOSPITAL

Top: A young woman happy to be photographed relaxing on a stretcher, book in hand and with her hair loose (HC198).

Bottom: Horse and trap ready and waiting by the gate at Polnish Hospital (HC22).

4: The Person behind the Camera

Above: Glen Mama Farm, near Polnish Hospital (HC087).

Right: Pets at play and on display, at Polnish Hospital. The dog to the bottom right features in the background of numerous photos from railway locations (HC007 and HC150-HC153).

A theory emerges

To sum up, our photographer visited the Mallaig Extension in February/ March 1900 and, as far as we know, did not form part of a delegation of official site visitors. This will have been an affluent, youthful and energetic horse-rider who enjoyed unhindered access to the railway works. He or she was on good terms with Robert McAlpine jnr. and had a particular fascination with the railway hospital at Polnish. These clues gave rise to a theory:

We knew that the third McAlpine brother, Thomas Malcolm, was the 'family photographer' (see p. 15), and that he was frequently described as a personable, inquisitive, energetic and well-liked youngster. He was also the person who came up with the contract-saving water turbine idea (ref. p. 45) and was put in charge of the Kinlochailort section of the line, featured in 109 of the photos. We also knew that he had every reason to be fascinated with the hospital at Polnish. Might he be our photographer, or perhaps our photographer's companion?

Thomas Malcolm McAlpine on horseback outside Polnish Hospital. In later life he became the owner of a racehorse stable. One of his horses even won the Grand National in 1921 (HC143).

Thomas Malcolm McAlpine at Polnish Hospital

On 6 May 1898 young Thomas Malcolm was supervising a blasting operation in a cutting near Beasdale when disaster struck. According to the notes made by PC McKay at Kinlochailort that day, Tom had been 'charging bores with gelignite' between Glenbeasdale and the head of Loch nan Uamh. Young and self-assured, he had failed to take sufficient cover and was hit on the left side of his body by a stone from one of the blasts. Severely injured, he was carried to the nearest navvy hut and the doctor fetched from Polnish. When Dr. Patrick arrived, he ordered the young contractor to be 'removed to the Cottage Hospital there in a conveyance.'

Dr. Patrick from Polnish ordered the injured man to be 'removed to the Cottage Hospital there in a conveyance'. Dogcarts were used as ad-hoc ambulances when required (HC230).

'… he was carried to the nearest navvy hut …'
Huts on the northern shores of Loch nan Uamh, below the entrance to the twin tunnels between Loch nan Uamh Viaduct and Glenbeasdale (HC081).

While Dr. Patrick, aged 27, would have been relatively inexperienced, he came from a family of medics and knew his limitations. He did his best, but Tom's pelvis was shattered and his ribs were crushed. A telegram was sent to the McAlpine family home in Old Kilpatrick:

'TOM STRUCK BY STONE. INTERNAL HAEMORRHAGES.
NOT EXPECTED TO LIVE.'

The cottage hospital at Polnish, with railway camp to the right (HC119).

4: The Person behind the Camera

Robert McAlpine snr. was a man of action. On receiving the devastating news, he immediately travelled to Glasgow's Charing Cross where the nation's most acclaimed surgeon, Professor William MacEwen, lived with his wife and children. By the evening, the 51-year-old pioneering railway contractor, nicknamed 'the Chief', had convinced the 50-year-old pioneering doctor, also nicknamed 'the Chief', to travel with him overnight to the West Highlands to see what could be done to save his son.

However, when Robert snr. tried to charter a train to take him north, the station official at Queen Street explained that despite the emergency, crossing Rannoch Moor at night would be impossible as no West Highland signalboxes were staffed at that hour of the day. As a compromise, McAlpine could charter a train to Craigendoran, just south of Helensburgh, where the train would have to wait overnight in readiness for an early departure north. But once they arrived at Craigendoran at half past ten that evening, Robert paid the train crew handsomely to break all the rules in the book and take the train across the moor, setting the points themselves as they went along. The anxious father gave his word that he would personally take the full responsibility were there to be any repercussions. [1]

In the many oral accounts of this story, the two famous gentlemen were the only passengers on this special overnight charter. But according to the *Inverness Courier*, they did in fact travel in the company of a specially trained nurse:

> Another Blasting Accident at the Mallaig Railway: [...] a doctor arrived by special train on Friday morning, accompanied by a trained nurse and Mr Robert McAlpine, senior partner of the firm. [...] (*Inverness Courier*, 10 May 1898)

A nurse in a position of authority, as indicated by her impressive bow. Her appearance tells us that this is not Polnish matron Belle Hawkins (HC201).

---

[1] MacEwen is also reported to have adopted the attitude of setting aside the rules of railway bureaucracy in the face of emergency. Having been refused an unscheduled stop near a location to which he had been urgently summoned, MacEwen once pulled the emergency cord of the express he was travelling on so he could alight where he was needed, handing the £5 fine to the guard as he alighted (Bowman, p. 323).

Professor William MacEwen (1848–1924) was born on the Isle of Bute and studied medicine in Glasgow. A student of Dr. Lister, he introduced systematic scrubbing, sterilisation of surgical tools, use of surgical gowns, and anaesthesia. He was one of the most innovative surgeons of his time and was able to greatly advance modern surgical technique and improve the recovery of patients. He was a pioneer in modern brain surgery and is considered the father of neurosurgery.

Working with pioneering nurse Rebecca Strong (1843–1944) in the 1880s, MacEwen began a training programme for nurses, focussing on sterilisation, at the Glasgow Royal Infirmary. He also encouraged his juniors and students to 'cultivate close relations' with the nurses who worked with them, 'for thereby much mutual good might accrue' (Bowman, p. 290).

Professor MacEwen enjoyed international acknowledgement and fame. He was knighted in 1902 for services to medicine.

A hotel coachman from Fort William took them west to Polnish (HC039).

Arriving in Fort William Station at 5am, the rescue party had to break out of the locked station. From there they proceeded to wake up a hotel coachman who, flustered by his early-morning surprise callers, answered the door in his top hat, shirt, overcoat and undergarments, much to the amusement of the two sleep-deprived 'Chiefs'. Embarrassed, he promptly agreed to take them west, and they arrived at Polnish seven hours later.

MacEwen operated immediately. Four days later there was still no sign of recovery and MacEwen declared that the patient would have to be transferred to Glasgow to receive proper care.

However, Thomas Malcolm was in a precarious state and could not travel in a carriage; the shaking would be sure to kill him. So, eight navvies stepped forward and volunteered to carry their boss on a stretcher. MacEwen gave them the training required, and so they set out on the gruelling journey.

The main road past Polnish Hospital (HC113).

4: *The Person behind the Camera*

The stretcher party's journey

Each stage of the stretcher party's journey is numbered to match the annotations on the maps in Appendix 1.

1. *From Polnish, they carried Thomas along the road high above Loch Dubh (HC102).*

2. *Then past Polnish Chapel and under the bridge known locally as 'the skew-whiff bridge'(HC027).*

3. *This took them down to Camas Driseach (HC051).*

4 – 6. *Then under the bridge opposite the bay and onwards up the brae along the Allt na Chriche burn, to the top of the bend where the railway stores and bakery stood (HC 101, HC054 and HC093).*

4: The Person behind the Camera

They headed downhill again and under the railway bridge just west of the Kinlochailort Hotel. By now they had walked a distance of some 3 miles.

This particular bend in the then very winding road, between Allt na Chriche and the hotel, disappeared from maps published after the 1905 revision.

7. *Kinochailort Hotel and post office (HC099).*

Robert snr. once had the reins of his horse stolen while having lunch here. They had found a new use holding up the trousers of a navvy.

The original hotel burnt down in 1994 but was soon rebuilt.

85

4: *The Person behind the Camera*

8. *Ailort Bridge at Arienskill (HC112).*

9. *Loch Eilt (HC123).*

They continued along the River Ailort to Ailort Bridge at Arienskill, where they walked under the railway and down to the western end of Loch Eilt. Here, seven boats were waiting so that the injured man, encased in plaster-of-Paris, could be carefully towed the four miles to the eastern end of the loch.

From the loch, the navvies continued along the Allt Lon a' Mhuidhe river up the steep two-mile climb to Lechavuie, then down the hill past the Bréin-choille navvy camp to Glenfinnan.

By the time they arrived in Glenfinnan, it was getting late and the navvies were exhausted. The local hotel, the Stagehouse (now the Prince's House Hotel), opened its doors to them, but it proved impossible to get the long stretcher through the tight corners in the entrance area. They resolved the problem by removing a window and carrying the young contractor through to the kitchen where the stretcher was placed on a long table overnight while the navvies got their rest. By then, they had taken their boss across a distance of some 13.5 miles (22 kilometres).

4: The Person behind the Camera

10. *The road along the Allt Lon a Mhuidhe river, on the climb up to Lechavuie from Loch Eilt (HC058).*

11. *The Stagehouse Inn, Glenfinnan – now the Prince's House Hotel. The kitchen window is on the ground floor, on the side wall (HC098).*

At the crack of dawn the next morning, they carried the young McAlpine yet another two miles past Glenfinnan Church and down to Loch Shiel, then on to the contractor's railway at Craigag, all the while making sure that the stretcher was kept perfectly still. The long chain coupling between the locomotive and the wagon with the stretcher posed a problem in this respect, as there would inevitably be jerking movements as the train went along. The navvies solved this by sitting in two rows opposite one another, legs touching, one row on the engine's buffer beam, the other on the frame of the wagon, their strong legs steadying the slow and painstaking ride. At Kinlocheil jetty the stretcher was carried onto a boat that took them to a waiting ambulance train. The navvies accompanied Thomas Malcolm to Glasgow and carried his stretcher through the streets for yet another mile from Queen Street Station to Rebecca Strong's private nursing home at no. 4 Queen's Crescent, where he later recovered.

We know nothing about how the navvies were rewarded for their extraordinary efforts. The McAlpines' private ledger does however include a payment to Professor MacEwen from Robert snr.'s account to the tune of £315, three times the annual rent he paid for his grand Dalnottar House. The two men remained friends for life.

Glenfinnan Church external bell (HC047); northern shores of Loch Shiel (HC105); contractor's locomotive and wagon (HC032).

It took the injured youngster some time to recuperate, but on 7 January 1899 he boarded a ship in Southampton with his 19-year-old sister Ethel and headed for Tenerife to convalesce in warmer climes. Their step-mother Florence joined them in the first week of February. Thomas Malcolm's private ledger reveals bills paid to hotels in Santa Cruz and Las Palmas until the end of March.

Is Thomas Malcolm our photographer?

After realising how well many of the photographs appear to document Thomas Malcolm's brush with death and his subsequent adventures in the West Highlands, we formulated the theory that on his return from Tenerife, the young McAlpine had travelled back to the Mallaig Extension in February 1900, camera in hand and with a companion, to revisit the scenes of his experiences and the people whose extraordinary efforts had saved his life.

We felt the most hopeful line of enquiry at this point would be to look for an undisputable connection between Thomas Malcolm and the photographs that were *not* from the Mallaig Extension. But how do you go about looking for the location of what appears to be random images of buildings, bridges and coastlines?

For who else was likely to have shown such keen interest in the people and pets at Polnish Hospital? Someone who worked there, perhaps? But then, would they dare approach Robert McAlpine jnr.'s residence with camera at the ready? And why would they photograph sections of road between Polnish and Glenfinnan? And who, in 1900, would be able to entice two hard-working nurses to be photographed while larking about on the ice, except a charming 22-year old whose life they had helped save?

We were dealing with a relatively wealthy individual who clearly took an interest in and had access to the railway works, yet the same person appeared to enjoy the company of nurses and was able to make an elderly man pose for the camera on a dogcart without a horse. All the other candidates we could think of, had to be discounted for one reason or another. Yet – while there were masses of circumstantial evidence, theories and hypotheses, there was no solid proof of the photographer's identity.

Polnish Hospital with railway camp to the right (HC119).

Ireland

Armed with the knowledge that Thomas Malcolm had been sent to Ireland in 1900 to supervise the building of the railway from Waterford to Fishguard, we started by looking for identifiable buildings in southern Ireland. But however many people's help we enlisted and however many searches we undertook in historical records, no clues materialised. It was Michael Holden who gave us our first lead. He pointed out that picture no. HC171 looked a lot like the gates of Camlin Castle, in County Donegal, Ireland.

Ireland's National Inventory of Architectural Heritage confirmed that HC171 was indeed of the Camlin Castle gates, so we changed tack and started looking to the north. After a few weeks of searching the National Library of Ireland's photographic collections, we found a match for picture no. HC174: the highly fashionable Great Northern Railway Hotel featured in an advert from 1899, promoting rail-borne tourism to the seaside resort of Bundoran. The luxury hotel had been built entirely from concrete, in 1894. And although the railway was long gone from Bundoran, the hotel was still there.

Once we knew we should be looking for photo locations along the route of the Great Northern Railway, things began to fall into place. Picture nos. HC170 – HC181, and a few others, were all touristy shots from County Donegal. We gradually identified the locations one by one, with the help of the National Library of Ireland's website and personal visits to Bundoran and County Donegal. We managed to obtain enthusiastic assistance from local museums and the current hotel management. We felt we were making good progress.

We found that all the pictures from the northern parts of Ireland were locations within easy reach of the railway between Dublin and Bundoran (built in 1866), and many of the photographed sites featured in railway advertisements of the times. We also established that according to the 1901 census, Thomas Malcolm and his younger brother Granville were staying in a swanky hotel in Dublin on 1 April. Fashionable Bundoran was only an express rail journey away.

River Erne, Ballyshannon (HC180), Falls of Erne, Ballyshannon (HC170)

The Fairy Bridges, Bundoran (HC176 & 181)

4: The Person behind the Camera

Young woman on a bicycle, outside the Great Northern Railway Hotel, Bundoran (HC160).

4: The Person behind the Camera

Whoever our photographer was, he or she clearly travelled by rail in Ireland as the 20th century dawned, looking at rivers, bridges and landscape formations. There was also a young woman on a bicycle outside the hotel in Bundoran who smiled happily at the photographer. But no-one was able to identify the lady, and the Great Northern Railway Hotel 's visitor records from the early 1900s had not been retained. There was therefore still no proof that Thomas Malcolm had been there – only evidence that he could have been.

We plodded on. Picture nos. HC193 and HC194 featured a couple of lighthouses. Surely, they would be possible to identify?

The lighthouse lead

We approached the Association of Lighthouse Keepers and showed the pictures to their archivist, asking where on the Scottish, English or Irish coast these lighthouses might have stood, sometime in the early 20th century.

The archivist came back with an answer almost straight away: We had been looking in all the wrong places, he said. These lighthouses were in fact off the coast of Portland, Maine – in the United States of America!

This opened up a whole new line of enquiry, and we sent some of the still unidentified seaside locations to the curator of Maine Maritime Museum. She was also quick off the mark: these pictures were all from Portland, with the Grand Trunk Railroad Grain Elevator being a key landmark.

Top: Portland Harbour lighthouse, near south Portland, Cumberland, USA (HC194).
Bottom: Portland Harbour pilot schooner (HC192).

Unidentified ship docked by the Grand Trunk Railroad Grain Elevator, Portland, Maine (HC202).

The SS Tunisian of Glasgow moored next to the SS Cervano of Dundee, in April 1900 or Jan, Feb or Mar 1901, photographed from a different ship (HC204).

4: The Person behind the Camera

Cargo ship moored by the Grand Trunk Railroad Grain Elevator, Portland Maine. Photographed from aboard the ship (HC208).

Transatlantic traveller with onlooking sailors. On close inspection, the sailors' hats read HMS on the rim, but the name of their ship is sadly illegible (HC205).

We now adjusted our search to focus on immigration and emigration records and found that the SS Tunisian, featured in picture nos. HC203 and HC204, had sailed from Liverpool to Portland on four occasions, in April 1900 and in January, February and March 1901.

We knew from the census records that Thomas Malcolm was in Dublin in April 1901 and Sir William had told us that his great-uncle had been criss-crossing the world throughout the early parts of the 20th century. Yet from 1899 to October 1901 he was the company's resident partner for their Leadhills & Wanlockhead railway contract in a remote area of Lanarkshire. There was clearly a bit of overlap between Scotland and Ireland when it came to Thomas Malcolm's established whereabouts during the period in question, which combined with the world-wide travel testimony had us reasonably convinced we were on the right track.

However, US immigration records eventually brought our search for proof to a disappointing end. Stating London as his final destination, Thomas Malcolm sailed from Cape Town via Sydney and Honolulu to San Francisco in December 1901/January 1902. On arrival in San Francisco he declared that this was his first time in the USA. So, if the records are correct, he could not have photographed the SS Tunisian in Portland the year before.

Admittedly, we have been unable to find any records that suggest how Thomas Malcolm travelled south to Cape Town, so in theory he might have sailed via Portland and stayed aboard a ship while there. However, we were now moving well into the realm of speculation, and felt we needed to draw a line under the lighthouse lead and declare that it had led us to a dead end.

The mystery medic

We had only one other lead left to research. Picture nos. HC167 and HC168 feature a pensive medical man, sitting by his desk in his study. He is approx. 50 years of age, is dressed in 'genteel country clothing', and has a lean build. He is resting his arm on a leather ledger and on his bookshelf there is a Catalogue of Surgical Instruments. There is a Buddha figurine on his mantelpiece and a microscope on his desk. The wall is adorned with a picture of Tower Bridge (built 1886 – 94) as well as anatomical drawings of a heart and a leg. The mirror reveals a telephone line running across the ceiling.

When trying to identify this man, we could immediately rule out the two young doctors who worked at Polnish, on account of their age. The other medic in the area at the time, Dr. Nicoll in Arisaig, was of the right maturity, but photographs of him revealed a different build.

For a long while we wondered whether this might in fact be Sir William MacEwen, the famous surgeon, who we knew had kept in touch with Thomas Malcolm after the Polnish drama, with Sir William having attended Thomas Malcolm's wedding to Maud Dees in 1903. When comparing our mystery medic to official portraits of MacEwen, there was however no definitive likeness, only a suggestive one.

After a round of consultations, we found that neither the archivist at Bute Museum nor the archivist at the Royal College of Surgeons and Physicians in Glasgow shared our hunch that the mystery medic might be a middle-aged MacEwen in his study on his country estate on the Isle of Bute. So, we decided to approach the portrait of our pensive doctor from a different angle.

4: The Person behind the Camera

Left: The mystery medic (HC168) ...
Above: ... and his capable assistant? (HC186)

So far we had been careful not to base any of our theories on the numbering sequence of the photographs, as Michael Holden had told us it was impossible to establish their right order based on the physical appearance of each delicate negative. Looking at their content, it was however clear that some of the locations depicted in consecutively numbered pictures related to one another, so much so that they formed a batch.

Pictures HC167 and HC168, which featured the mystery medic, were wedged between a series of photos from Bundoran and neighbouring Ballyshannon in County Donegal, as was picture no. HC186, which featured the smiling nurse from picture no. HC201 (see p. 82), now dressed as if for travel.

4: The Person behind the Camera

A cottage hospital maternity ward (HC 185).

We asked ourselves the question:

Did MacEwen's trained theatre nurse (see p. 81), who accompanied the famous doctor to Polnish, later move on to a cottage hospital, perhaps in Donegal, as assistant to the mystery medic in pictures HC168 and HC167?

When we first looked through the photos and came across picture no. HC185, we had excitedly wondered whether this might be an internal shot from Polnish but soon discovered that the layout of the room (fireplace and windows) did not match that of Polnish House. We now looked closer at this particular image for further clues and realised we had missed something that had been staring us in the face:

This is a picture of a hospital ward for women and children! There are empty baby cots in the foreground on both sides of the aisle, and there are women's shawls hanging over several bed-ends. An online search told us that Ballyshannon District Hospital, which first opened for the local workhouse in 1843, did indeed have a 'small operation ward and maternity ward' (Survey of Hospital Archives in Ireland, p. 24), so this was at least a possibility.

4: The Person behind the Camera

This picture clearly has importance in a wider context than the Mallaig Railway. We promptly sent it to the Centre for the History of Medicine in Ireland at University College Dublin, who we felt would be in a better position to fully appreciate its story.

In the context of our search for the identity of our photographer, we had however learnt a few important things from our detours to Portland and Donegal:

The photographer had criss-crossed the globe at the very beginning of the 20th century, just like Sir William McAlpine had told us that Thomas Malcolm had been doing. Furthermore, railways run as the only common denominator between the various regions featured: the Mallaig Extension in the Scottish West Highlands, the Great Northern Railway in Ireland, and the Grand Trunk Railroad that runs from Portland, Maine to Montreal, Canada.

An educated guess

Despite all our painstaking work to establish the identity of our photographer, we have eventually had to accept that we may never know for certain who took the pictures. However, we set out to learn what might have motivated the pointing of the lens, and that, I think, we have achieved.

We know the pictures were taken by a wealthy individual involved with the building of the Mallaig Railway, who travelled round the world and who related well to people from all walks of life. We also know that the pictures document Thomas Malcolm's brush with death in the West Highlands, and that the person behind the camera goes to great lengths to honour with a photograph some of the individuals who played a part in his rescue.

It is our educated guess that Thomas Malcolm McAlpine, who changed his first name to Malcolm in 1901, visited the West Highlands with a companion in February 1900 to revisit the sites involved with his accident and rescue, and to thank the people involved. They then went on to seek out the trained theatre nurse who had assisted Professor MacEwen during the life-saving operation on the young railway contractor and who may well have tended to him during his recuperation in Glasgow. She may even be 'the girl on a bicycle' outside the Great Northern Railway Hotel on p. 92.

Whether the owner of the camera was Thomas Malcolm himself, or a travel companion – be it friend or family, we do know that the point of view reflected in these photographs is inspired by his story. Whoever the photographer was, we feel we have revealed their perspective. We hope, however, that this book has also served to widen that perspective by casting a light on some of the Mallaig Railway stories of people far less fortunate.

It is our hope that by explaining our research process in detail, including all the blind alleys, interested readers will feel inspired to pick up the thread where we left it. If you have any further leads, answers or ideas, we would love to hear from you. The full Holden Collection is shown as thumbnails in Appendix 5.

The high-resolution images can be viewed at Glenfinnan Station Museum for free. A license can be obtained for using the photos, and prints of many of the images are available for sale from the museum website: www.glenfinnanstationmuseum.co.uk.

4: The Person behind the Camera

Thomas Malcolm McAlpine posing in the snow outside Polnish Hospital (HC141).

Matron Belle Hawkins posing in the snow outside Polnish hospital (HC139).

Polnish hospital staff nurse posing in the snow outside the hospital (HC140).

4: The Person behind the Camera

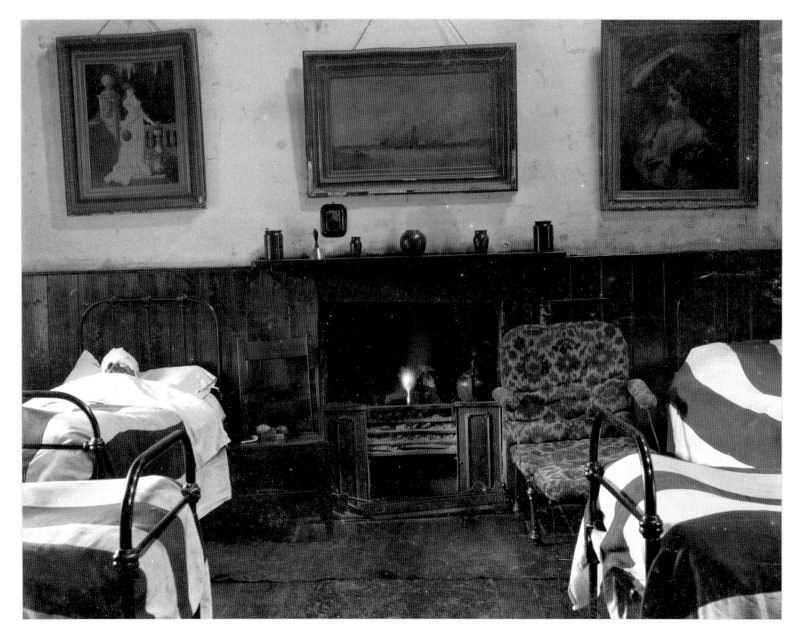

Injured male patient in a cottage hospital, 1900 (HC182).

Appendix 1 – Maps

This appendix provides extracts from a series of Ordnance Survey maps published in the period 1900 – 1908. Together, they illustrate the lie of the land surrounding the Mallaig Railway at the beginning of the 20th century. The spelling chosen for the Gaelic place names in the main chapters reflects the spelling found on the maps in this appendix. The spelling chosen for railway structures reflects the spelling found in the contractor's records.

We have used different map editions for different sections of the line, in the interest of providing as accurate and informative an illustration as possible of the points made, stories told and locations identified in the main chapters. This also gives you an opportunity to follow Thomas Malcolm McAlpine's stretcher journey in detail, matching the numbered photos in Chapter 4 to specific map locations.

Each map is accompanied by its OS edition number as well as its year of survey and year of publishing.

ALL MAPS ARE REPRODUCED
WITH THE PERMISSION OF
THE NATIONAL LIBRARY OF SCOTLAND

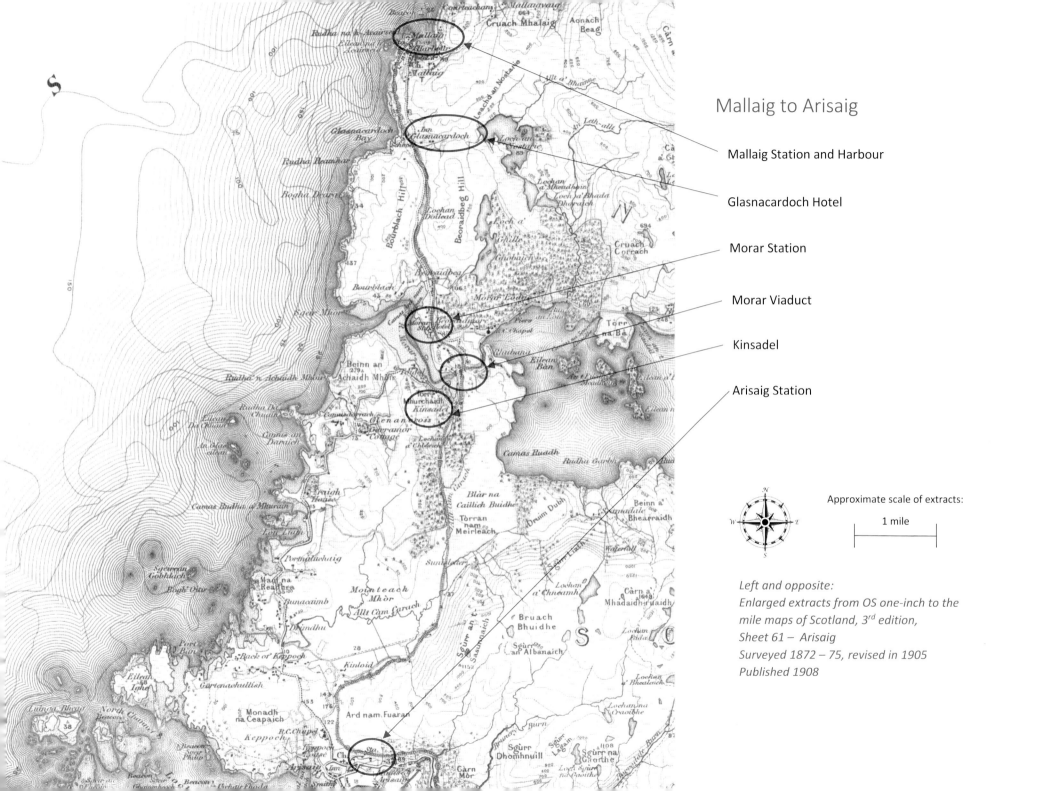

Mallaig to Arisaig

Mallaig Station and Harbour

Glasnacardoch Hotel

Morar Station

Morar Viaduct

Kinsadel

Arisaig Station

Approximate scale of extracts:

1 mile

Left and opposite:
Enlarged extracts from OS one-inch to the mile maps of Scotland, 3rd edition,
Sheet 61 – Arisaig
Surveyed 1872 – 75, revised in 1905
Published 1908

Appendix 1: Maps

Arisaig to Lochailort

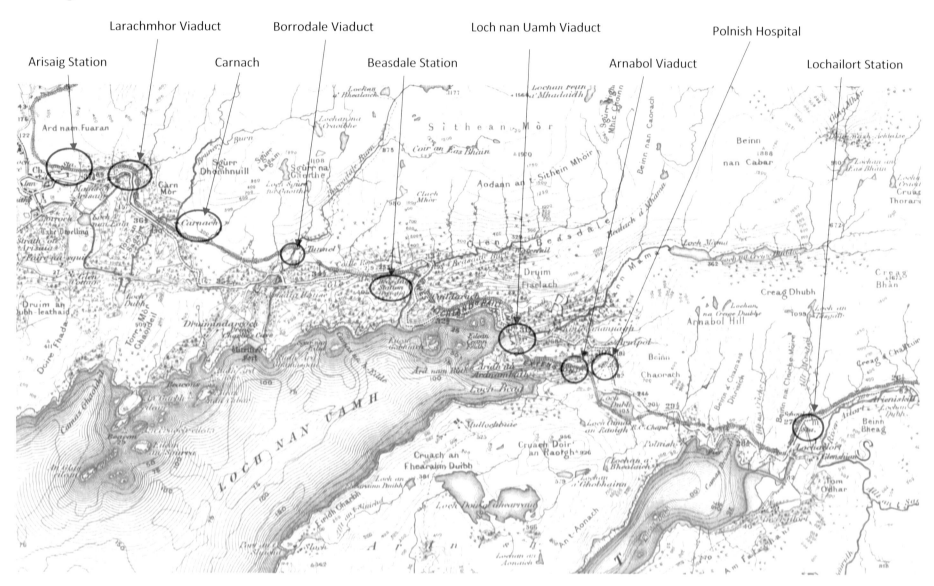

103

Polnish and Kinlochailort – a closer look

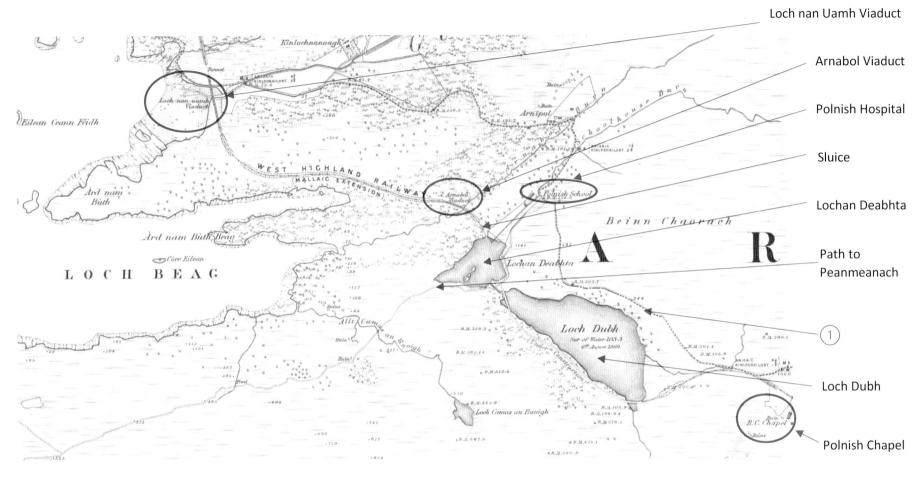

- Loch nan Uamh Viaduct
- Arnabol Viaduct
- Polnish Hospital
- Sluice
- Lochan Deabhta
- Path to Peanmeanach
- Loch Dubh
- Polnish Chapel

Above:
Enlarged extract from OS six-inch to the mile map,
2nd edition, Inverness-shire - Mainland Sheet CXXXV
Surveyed 1873, revised in 1899
Published 1900

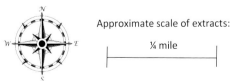

Approximate scale of extracts: ¼ mile

Opposite:
Enlarged extract from OS six-inch to the mile map,
2nd edition, Inverness-shire - Mainland Sheet CXXXVI
Surveyed 1873, revised in 1899
Published 1902

Appendix 1: Maps

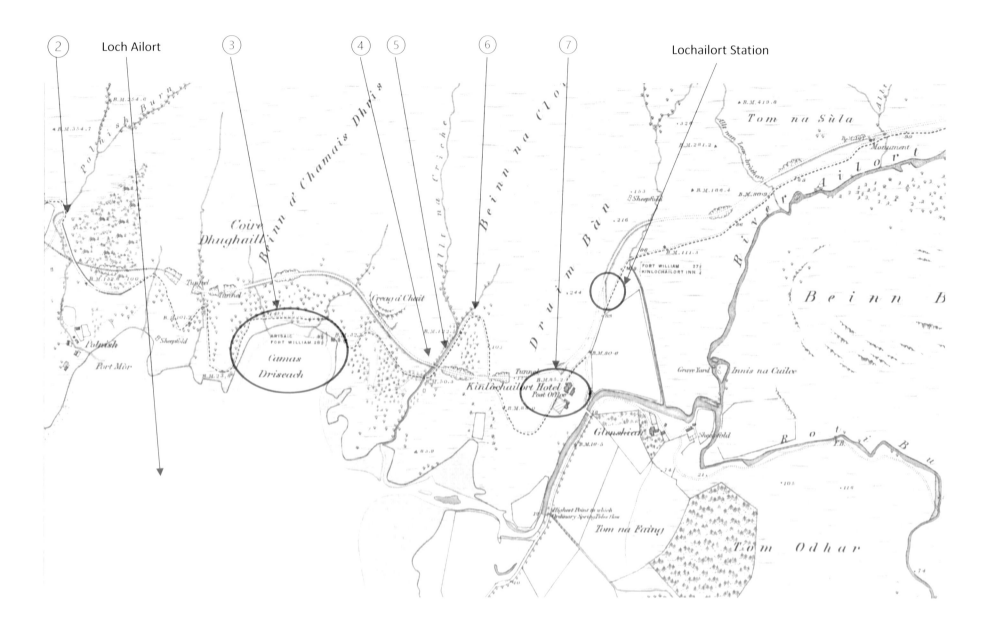

Arienskil and Loch Eilt – a closer look

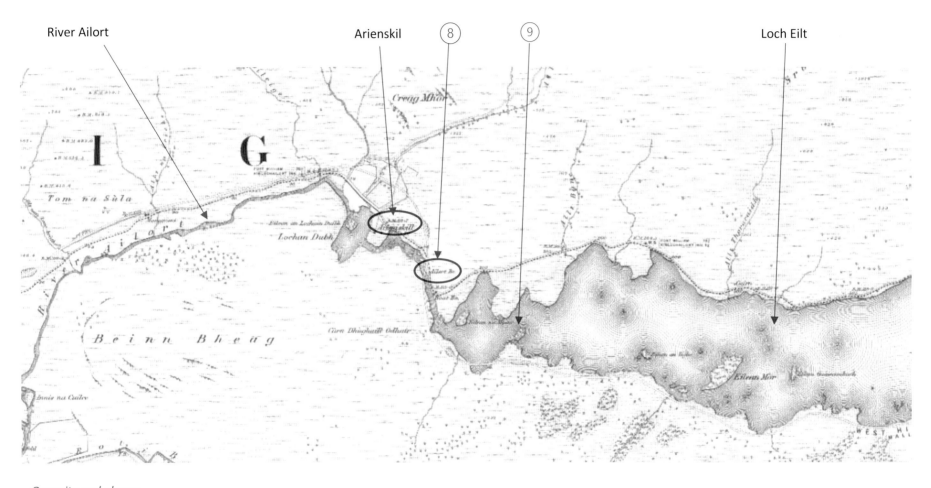

Opposite and above:
Extracts from OS six-inch to the mile map, 2nd edition
Inverness-shire - Mainland Sheet CXXXVI
Surveyed 1873, revised in 1899
Published 1902

Approximate scale of extracts:

¼ mile

Appendix 1: Maps

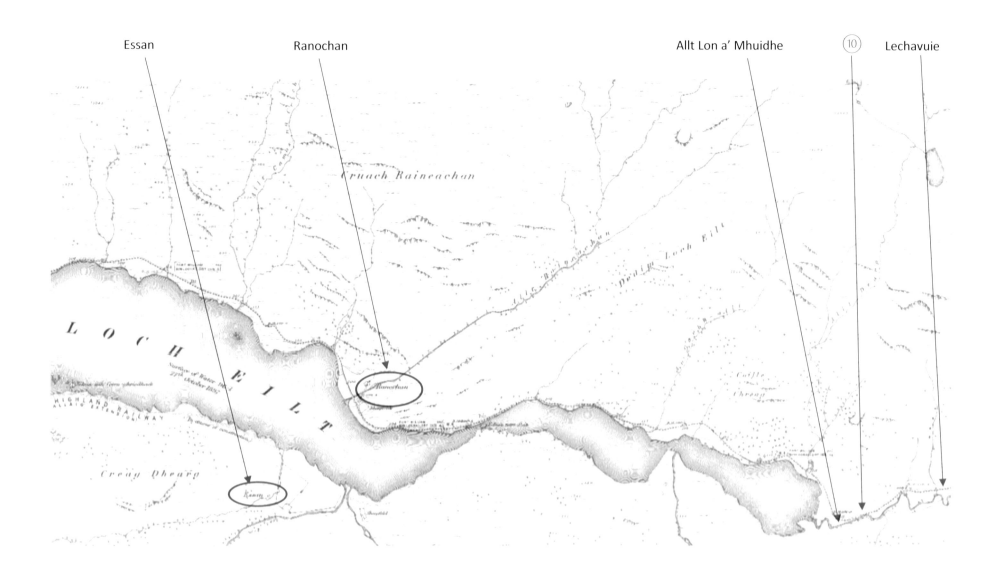

Glenfinnan to Banavie

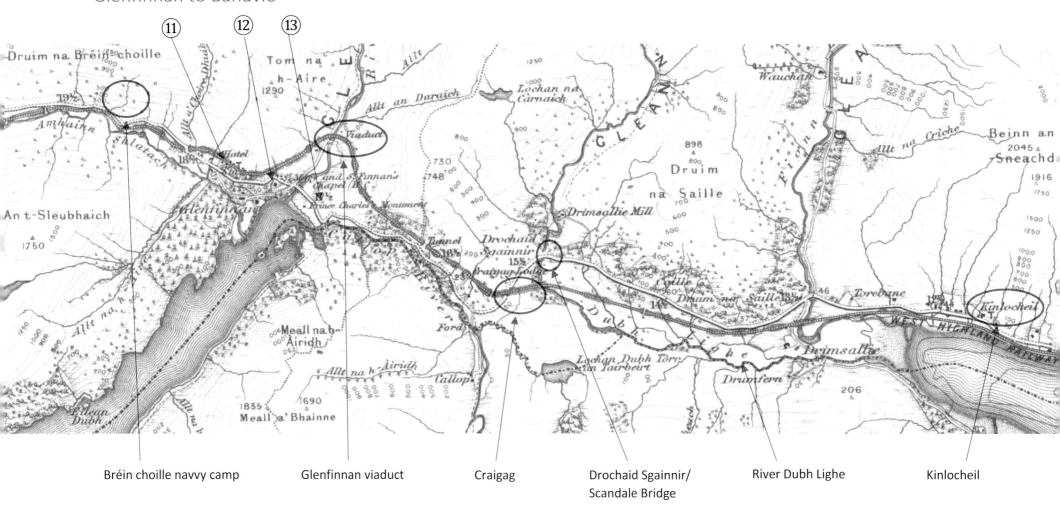

Bréin choille navvy camp · Glenfinnan viaduct · Craigag · Drochaid Sgainnir/Scandale Bridge · River Dubh Lighe · Kinlocheil

Above and opposite:

Extracts from OS one-inch to the mile maps of Scotland, 3rd edition, Sheet 62 – Loch Eil
Surveyed 1868 – 72, revised in 1905, published 1908

Appendix 1: Maps

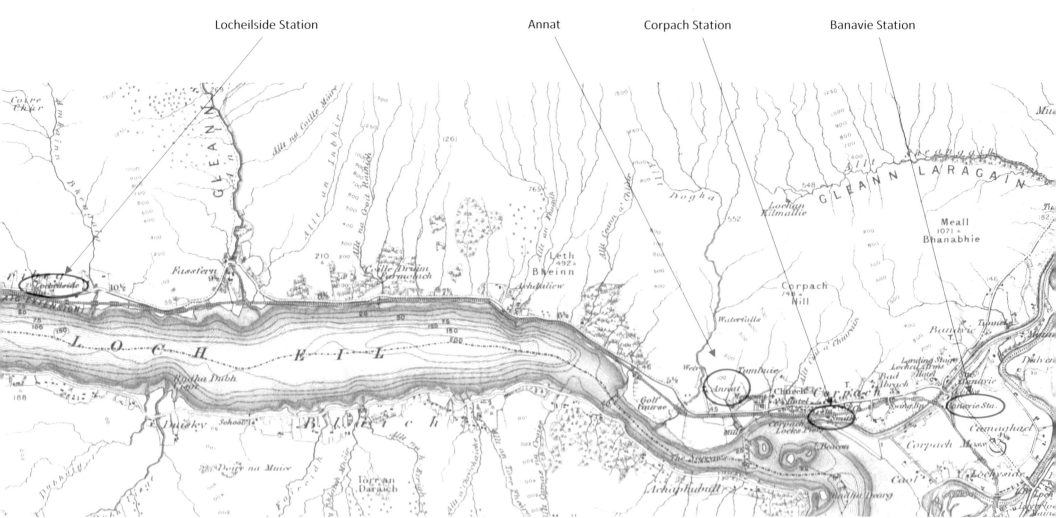

Appendix 2 – Locomotives

To our knowledge, the most comprehensive list of locomotives known to have been used by Robert McAlpine & Sons on the Mallaig Extension has been compiled by the Industrial Locomotive Society. In 1990 they published a well-researched booklet with details of locomotives used on contracts in Scotland. This lists ten engines that were definitely used by the McAlpines on the Mallaig Railway, plus one that was considered a 'may-be'. All of these were either 0-4-0 or 0-6-0 saddle tanks.

The Holden Collection provides a detailed photographic record of four of the eleven locomotives listed by the Industrial Locomotive Society, including the 'may-be' one, here featured on p. 112, thereby moving it into the 'definite' category.

The locomotive on the page opposite is the oldest engine featured in the Holden Collection. This 0-6-0 saddle tank was built as no. 139 by Fox Walker & Co in Bristol in 1871. (Fox Walker & Co was taken over by Peckett and Sons in 1881).

According to the Industrial Locomotive Society, it was sold new to contractors Lucas & Aird who gave it the running number 138 when it became a part of their fleet. Eighteen years later they used it for their West Highland Line contract to Fort William and Banavie. When they gave up waiting for the Mallaig Extension to receive its parliamentary go-ahead (see p. 11), they may well have left it behind in the West Highlands. For here, in 1900, we find this tired-looking workhorse just east of Arisaig, in the employment of Messrs. Robert McAlpine & Sons.

You may recognise the driver/ fireman team on the footplate, believed to be father & son, featured on p. 21. The location is Carnach, by Larachmhor Viaduct.

Note the wooden 'dumb buffers'. The white-ended barrel lying trackside is marked 'Engine Oil'.

Opposite: Fox Walker 0-6-0 saddle tank locomotive previously owned by Lucas & Aird & Co (HC018).

Appendix 2: The locomotives

Hudswell Clarke & Co. no. 532, Waverley (HC017).

For use on their various contracts, Robert McAlpine & Sons bought ten new steam locomotives from Hudswell Clarke of Leeds in 1898/99, each costing between £750 and £850 (Russell p. 89).

This is one of them: 'Waverley', an 0-4-0 saddle tank with an open cab.

The maker's plate reads:
 'Hudswell Clarke & Co. Railway Foundry. No. 532, LEEDS, 1899. '

The Industrial Locomotive Society's archivist has told us that 'Waverley' is recorded as sent to Glasgow for onward shipment, so it will most likely have arrived in the West Highlands by sea. It was later used in Ellesmere Port at McAlpine's Great Stanney Depot.

The locomotive is parked over a pit (note the person working underneath its front buffers) at a location which has yet to be identified.

Appendix 2: The locomotives

This is another Leeds-built engine from Hudswell Clarke Railway Foundry. Sadly, the maker's plate has caught the sun so is difficult to read, but this is most likely no. 492, built in 1898, which is known to have been used by the McAlpines for the Mallaig Extension contract (ILS, 1990). See p. 44 for a similar locomotive at work.

No. 492 was later used by the McAlpines for their Provan Gas Works contract.

Note the sprung buffers and enclosed cab, providing added protection against the weather.

A Leeds built 0-6-0ST locomotive, believed to be no. 492, posing at Bridge 66 by Glenfinnan Station (HC036).

Appendix 2: The locomotives

North British Railway locomotive no. 895 at Arisaig Station (HC020). Hired from NBR and returned (ILS, 1990).

This is North British Railway Locomotive 895 – a Class Y9 Holmes saddle tank specifically designed for shunting. It would normally be used for working on the docks or in other industrial settings. It is therefore surprising to see it putting in an appearance during the building of the Mallaig Railway.

It was built by Neilson & Co. at Cowlairs in 1891, then as no. 50. It was renumbered 895 in 1897. The North British Study Group's archivist has told us that the loco was later renumbered again, to 1095, presumably in 1911 when the 4-4-0 no. 895 was built. The locomotive became LNER 10095 in 1923 and LNER 8111 in 1946. It was withdrawn in 1953 carrying BR No. 68111.

Here photographed at Arisaig Station, note the newly built platforms, discarded planks from the concrete formwork and the tip of the overhanging eaves of the station building, top left.

Appendix 3 – The Turbines

In 1908 John Wilton Cuninghame Haldane published *Railway Engineering, Mechanical and Electrical* simultaneously in London and in New York. The book is now out with copyright and is considered culturally important and part of the knowledge base of civilization as we know it. The book has therefore been digitised by Google from the library of the University of Michigan and uploaded to the Internet Archive.

The Internet Archive is buildig a digital library of Internet sites and other cultural artifacts in digital form, providing free access to researchers, historians, scholars, the print-disabled, and the general public. Their mission is to provide Universal Access to All Knowledge. Books published prior to 1923 are available for download.

The following is a quote from Chapter 28, 'Primary Aids to Success in Railway Enterprises':

DOUBLE DISCHARGE HORIZONTAL TURBINE

The engraving shows the exterior of a turbine, ready for setting in a masonry or brickwork well, the main vertical shaft transmitting power, the small one being used for regulating the admission of water to the interior. When of single or duplex horizontal formation, the turbine may be placed on a bed plate upon the ground level, and cased in with plate iron, the pipes being similarly made, all the parts being most convenient for practical use, and easily accessible for examination or repair.

The Double Discharge Horizontal Turbine of Messrs. Carrick & Ritchie, for falls of 10' 0" to 50' 0", with its starting and governing appliances, is shown on this page. Here, for example, two 9" wheels with 50' 0" head of water and 955 revolutions per minute, produce a power of about 42 horses either for belt or for wheel gearing (Haldane 1908, p. 520).

The quote is accompanied by this illustration:

It was Edinburgh-based hydraulic engineers Messrs. Carrick & Ritchie that supplied Robert McAlpine & Sons with double discharge horizontal turbines for the Mallaig Extension.

© National Railway Museum / Science and Society Picture Library

Appendix 4 – National Railway Museum Photographs

The National Railway Museum in York have a number of photographs in its collection relating to the Mallaig Extension to the West Highland Line. These are held in three different albums credited to George MacLachlan, who is said to have been an engineer working on the line. They include images from an earlier stage of the building works than the Holden Collection, and therefore complement our own photographs. To provide as complete a picture as possible of the construction process, we therefore felt it would be important to include a few of them here.

Some of the most notable pictures from the NRM collection (including those shown in this appendix) were reproduced in Broadsheet no. 10 'The Mallaig Railway', published by the Royal Commission on the Ancient and Historical Monuments of Scotland (RCAHMS) in 2002. RCAHMS merged with Historic Scotland in 2015 to form Historic Environment Scotland. The Broadsheet is now sadly out of print.

Opposite:
Arnabol Viaduct under construction (NRM10620540). The formwork for the arches is in place and the abutments & foundations have been completed. This dates the photograph to the late summer of 1898. The tent roof in the foreground is assumed to provide temporary cover for the turbine described in the previous appendix.

Right:
Rock crusher operated by stationary steam engine (NRM10620528). Rocks were dropped into the mechanical crusher positioned on the raised platform and crushed between metal plates to the desires size of aggregate. The noise would have been deafening.

Appendix 4: The National Railway Museum's Photographs

Loch nan Uamh Viaduct (NRM10620528).
The westernmost end of the viaduct looks fully completed while the formwork is still up for the four easternmost arches, see p. 56.

Appendix 5 – The Holden Collection as thumbnails

Appendix 5: The Holden Collection as thumbnails

Appendix 5: The Holden Collection as thumbnails

Appendix 5: The Holden Collection as thumbnails

Appendix 5: The Holden Collection as thumbnails

Appendix 5: The Holden Collection as thumbnails

Appendix 5: The Holden Collection as thumbnails

Appendix 5: The Holden Collection as thumbnails

126

Appendix 5: The Holden Collection as thumbnails

Appendix 5: The Holden Collection as thumbnails

Bibliography

Bowman, A.K. (1942). *The Life and Teaching of Sir William MacEwen; a Chapter in the History of Surgery.* London: William Hodge and Company

Glenfinnan Police Station, *Daily journal of duties 1897 – 1901.* Lochaber Archive Centre R91/DC/5/12/1-3

Grant, Ogilvie (1891). *Annual Report of the Medical Officer.* Inverness-shire County Council

Haldane, J. W.C. (1908). *Railway Engineering, Mechanical and Electrical.* London: E & F.N Spon Ltd.

Hardie, A.M. (c. 1960). *The Story of Robert McAlpine and his Family.* Unpublished. University of Glasgow Archive Services. GB 248 UGD 254/22/2

Heber, S. (1924). *Da Bergensbanen blev til.* Kristiania: Gyldendalske Bokhandel

Kinlochailort Police Station, *Daily journal of duties 1897 – 1901.* Lochaber Archive Centre R91/DC/5/14/1-4

Lund, S.A. (1900). 'Gravhalstunnelen', *Teknisk Ugeblad* (vol. 47), pp. 641-652

MacKenzie, C. (c. 1940). *McAlpine.* Unpublished National Library of Scotland, ACC. 13252

Mason Photographic Collection M20/33-35. Dublin: National Library of Ireland

McGregor, J. (2005). *The West Highland Railway: Plans, Politics and People.* Edinburgh: Birlinn Ltd.

McGregor, J. (2013). *The West Highland Extension.* Stroud: Amberley Publishing

McGregor, J. (2015). *The New Railway.* Stroud: Amberley Publishing

National Archives of Ireland (2015). *Survey of Hospital Archives in Ireland.* Dublin: National Archives of Ireland

Pringle, I.W. (1901). Inspection Report. National Archives, Kew MT6/1034/2

Report of H.M. Inspector 1897 - 1900, Public School Glenfinnan. Lochaber Archive Centre CJ/1/5/3/1429

Russell, I. F (1986). *Sir Robert McAlpine & Sons – the early years.* Carnforth: Parthenon Publishing

Shipway, J. S. (1998) . 'The Making of the West Highand Railway 1889 – 1901'. *Transactions*, vol. 141 (1997-98), paper no. 1565. Glasgow: The Institution of Engineers and Shipbuilders

Sir Robert McAlpine & Sons. Ledgers and Contract Papers (1897-1901). *Records of Sir Robert McAlpine & Sons Ltd, civil engineering contractors, 1882-1982.* University of Glasgow Archive Services. GB 248 UGD 254/1-5

Sir Robert McAlpine & Sons (2001). *The Mallaig Railway. The West Highland 1897 – 1901.* Edinburgh: Royal Commission on the Ancient and Historical Monuments of Scotland (RCAHMS)

Thomas, J. (1965). *The West Highland Railway.* Devon: David St. John Thomas

Tucker, D.G. (1976). 'Hydroelectricity for public supply in Britain, 1881 – 1894'. *Proceedings of the Institution of Electrical Engineers,* Vol. 123 (10), pp. 1026 – 1034. IET Digital Library

Tyrell, H.G (1909). *Concrete Bridges and Culverts, for both Railroads and Highways.* London: E & F.N Spon Ltd

Unknown (1898). 'The Mallaig Railway'. *The Engineer*, 16 Sept 1898, pp. 273-275

Wilson, W.S. (1907). 'Some Concrete Viaducts on the West Highland Railway'. *Minutes of the Proceedings of the Institution of Civil Engineers.* Vol. 170, paper no. 3595, pp. 304 – 307. ICE Publishing